Encyclopaedia of
DNA Recombinant Technology

Encyclopaedia of
DNA Recombinant Technology

Volume 3

Editor:
R.N. Kumar

AGROTECH PRESS
JAIPUR ▪ NEW DELHI

AGROTECH PRESS

Regd. Off. "Bony Residency", Gate No. 2
Opp. Tilak Public School, Vishweswariya Nagar,
Gopalpura Road, Jaipur - 302018, India
Ph: 0141 - 2761280, Telefax: 2761381
E-mail : agrotechpress@gmail.com

Delhi Office: 102, 1st Floor, Satyam House,
4327/3, Ansari Road, Darya Ganj,
New Delhi - 110002, India
Ph: 011-45652440
visit us: www.agrotechpress.com

Distributed By
KEYA INTERNATIONAL PRESS LLP
E-mail: keyainternationalpressllp@gmail.com

Published in India, 2024

ISBN : 9789384568214 (3 Vols. Set)

Typeset by Thomson Press

Printed and bound in India by Thomson Press Ltd.,
New Delhi

Contents

Preface

rDNA stands for recombinant DNA. DNA is the keeper of the all the information needed to recreate an organism. All DNA is made up of a base consisting of sugar, phosphate and one nitrogen base. There are four nitrogen bases, adenine (A), thymine (T), guanine (G), and cytosine (C). The nitrogen bases are found in pairs, with A & T and G & C paired together. The sequence of the nitrogen bases can be arranged in an infinite ways, and their structure is known as the famous "double helix" which is shown in the image below. The sugar used in DNA is deoxyribose. The four nitrogen bases are the same for all organisms. The sequence and number of bases is what creates diversity. DNA does not actually make the organism, it only makes proteins. The DNA is transcribed into mRNA and mRNA is translated into protein, and the protein then forms the organism. By changing the DNA sequence, the way in which the protein is formed changes. This leads to either a different protein, or an inactive protein.

Recombinant DNA is the general name for taking a piece of one DNA, and combining it with another strand of DNA. Thus, the name recombinant! Recombinant DNA is also sometimes referred to as "chimera." By combining two or more different strands of DNA, scientists are able to create a new strand of DNA. The most common recombinant process involves combining the DNA of two different organisms. There are three different methods by which Recombinant DNA is made. They are Transformation, Phage Introduction, and Non-Bacterial Transformation. Recombinant DNA works when the host cell expresses protein from the recombinant genes. A significant amount of recombinant protein will not be produced by the host unless expression factors are added. Protein expression depends upon the gene being surrounded by a collection of signals which provide instructions for the transcription and translation of the gene by the cell. These signals include the promoter, the ribosome binding site, and the terminator. Expression vectors, in which the foreign DNA is inserted, contain these signals. Signals are species specific. In the case of E. Coli, these signals must be E. Coli signals as E. Coli is unlikely to understand the signals of human promoters and terminators. Recombinant Problems are encountered if the gene contains introns or contains signals which act as terminators to a bacterial host. This results in premature termination, and the recombinant protein may not be processed correctly, be folded correctly, or may even be degraded. Production of recombinant proteins in eukaryotic systems generally takes place in yeast and filamentous fungi. The use of animal cells is difficult due to the fact that many need a solid support surface, unlike bacteria, and have complex growth needs. However, some proteins are too complex to be produced in bacterium, so eukaryotic cells must be used. DNA has been gaining in importance over the last few years, and recombinant DNA will only become more important in the 21st century as genetic diseases become more prevelant and agricultural area is reduced. Below are some of the areas where Recombinant DNA will have an impact. Scientists associated with the initial development of recombinant DNA methods recognized that the potential existed for organisms containing recombinant DNA to have undesirable or dangerous properties. At the 1975 Asilomar Conference on Recombinant DNA, these concerns were discussed and a voluntary moratorium on recombinant DNA research was initiated for experiments that were thought to be particularly risky. This moratorium was widely observed until the National Institutes of Health (USA) developed and issued formal guidelines for rDNA work. Today, recombinant DNA molecules and recombinant proteins are usually not regarded as dangerous. However, concerns remain about some organisms that express recombinant DNA, particularly when they leave the laboratory and are introduced into the environment or food chain. These concerns are discussed in the articles on genetically-modified organisms and genetically-modified food controversies.

The present textbook on DNA Recombinant Technology has been designed for university students, scholars and teachers dealing with genetics, molecular biology, genetic engineering, gene cloning, gene manipulation, microbiology and biochemistry. The overarching theme of the book the important link between genes and genetic engineering, is mirrored in the text. Efforts have been made to present the principles of recombinant technology, and its associated techniques, in sufficient detail to enable the non-specialist reader to understand them.

—*Editor*

1

Regulation of Gene Expression in Eukaryotes

The cloning of Dolly, a sheep, was reported worldwide in 1996. Dolly developed from adult somatic nuclei that had been implanted into enucleated eggs (eggs with the nuclei removed). More recently, cows, pigs, mice, and other mammals have been cloned as well with the use of similar technology. The successful cloning of Dolly was a great surprise to the scientific community because the cloning of mammals from somatic cells was thought to be impossible.

A reason for the initial skepticism was that the formation of male and female gametes (sperm and egg cells) was known to include sex-specific modifications to the respective genomes that resulted in sex-specific patterns of gene expression. As such, Dolly is symbolic of how far we have progressed in understanding aspects of eukaryotic gene regulation such as the global control of gene expression exemplified by gamete development. However, for every successful clone, including Dolly, there are many more, perhaps hundreds of embryos that fail to develop into viable progeny. The extremely high failure rate underscores how much remains to be deciphered about eukaryotic gene regulation.

The biological properties of each eukaryotic cell type are largely determined by the proteins expressed within it. This constellation of expressed proteins determines much of the cell's architecture, its enzymatic activities, its interactions with its environment, and many other physiological properties. However, at any given time in a cell's life history, only a fraction of the RNAs and proteins encoded in its genome are expressed. At different times, the profile of expressed gene products can differ dramatically, both in regard to which proteins are expressed and at what levels. How are these specific profiles generated?

As one might expect, if the final product is a protein, regulation could be achieved by controlling the transcription of DNA into RNA or the translation of RNA into protein. In fact, gene regulation takes place at many levels, including at the mRNA level (through alterations in splicing or the stability of the mRNA) and after translation (by modifications of proteins). However, most regulation is thought to take place at the level of gene transcription; so, in this chapter, the primary focus is on the regulation of transcription.

The basic mechanism at work is that molecular signals from outside or inside the cell lead to the binding of regulatory proteins to specific DNA sites outside of protein-encoding regions, and the binding of these proteins modulates the rate of transcription. These proteins may directly or indirectly assist RNA polymerase in binding to its transcription initiation site—the *promoter*—or they may repress transcription by preventing the binding of RNA polymerase.

Although bacteria and eukaryotes have much of the logic of gene regulation in common, there are some fundamental differences in the underlying mechanisms and machinery. Both use sequence-specific DNA-binding proteins to modulate the level of transcription. However, eukaryotic genomes are bigger and their range of properties is larger than those of bacteria. Inevitably their regulation is more complex, requiring more types of regulatory proteins and more types of interactionswith the adjacent regulatory regions in DNA. The most important difference is that eukaryotic DNA is packaged into *nucleosomes,* forming *chromatin,* whereas bacterial DNA lacks nucleosomes. In eukaryotes, chromatin structure is dynamic and is an essential ingredient in gene regulation.

In general, the ground state of a bacterial gene is "on." Thus, RNA polymerase can usually bind to a promoter when no other regulatory proteins are around to bind to the DNA. In bacteria, transcription initiation is prevented or reduced if the binding of RNA polymerase is blocked, usually through the binding of a repressor regulatory protein. Activator regulatory proteins increase the binding of RNA polymerase to promoters where a little help is needed. In contrast, the ground state in eukaryotes is "off."

Therefore, the transcriptional machinery (including RNA polymerase II and associated general transcription factors) cannot bind to the promoter in the absence of other regulatory proteins. In many cases, the binding of the transcriptional apparatus is not possible, owing to the position of nucleosomes near the promoter. Thus, chromatin structure usually has to be changed to activate eukaryotic transcription. The structure of chromatin around activated or repressed genes within cells can be quite stable and inherited by daughter cells. The inheritance of chromatin structure is a form of inheritance that does not directly entail DNA sequence.

1. In bacteria, all genes are transcribed into RNA by the same RNA polymerase, whereas three RNA polymerases function in eukaryotes.
2. RNA transcripts are extensively processed during transcription in eukaryotes; the 5_ and 3_ ends are modified and introns are spliced out.
3. RNA polymerase II is much larger and more complex than its bacterial counterpart. One reason for the added complexity is that RNA polymerase II must synthesize RNA *and* coordinate the special processing events unique to eukaryotes.

Multicellular eukaryotes may have as many as 25,000 genes, severalfold more than the average bacterium. Moreover, patterns of eukaryotic gene expression can be extraordinarily complex. That is, there is great variation among genes in when a gene is on (transcribed) or off (not transcribed) and in how much transcript needs to be made. For example, one gene may be transcribed only during early development and another only in the presence of a viral infection. Finally, the majority of the genes in a eukaryotic cell are off at any one time. On the basis of these considerations alone, eukaryotic gene regulation must be able to ensure that the expression of most genes in the genome is off at any one time while activating a subset of genes; and generate thousands of patterns of gene expression.

Mechanisms have evolved to ensure that most of the genes in a eukaryotic cell are not transcribed. Before considering how genes are kept transcriptionally inactive, we will focus on the second point: How are eukaryotic genes able to exhibit an enormous number and diversity of expression patterns? The machinery required for generating so many patterns of gene transcription in vivo has many components, including both regulatory proteins and cis-acting regulatory sequences. To initiate transcription, these proteins interact with DNA sequences called promoter-proximal elements near the promoter of a gene. The second group of protein components consists of specific transcription factors that bind to cis-acting regulatory sequences in the DNA called enhancers or upstream activating sequences (UAS's). These regulatory sequences may be located a considerable distance from gene promoters. Generally speaking, promoters and promoter-proximal elements are bound by transcription factors that affect the expression of many genes. Enhancers are the targets of more specific transcription factors that control the regulation of smaller subsets of genes. Often, an enhancer will act in only one or a few cell types in a multicellular eukaryote. For RNA polymerase II to transcribe DNA into RNA at a maximum rate, multiple cis-acting regulatory elements must play a part.

The promoters, promoter-proximal elements,and enhancers are all targets for binding by different trans-acting DNA binding proteins. The binding of RNA polymerase II to the promoter does not produce efficient transcription by itself.

Transcription requires the binding of general transcription factors to additional promoter-proximal elements that are commonly found within 100 bp of the transcription initiation site of many (but not all) genes. One of these elements is the CCAAT box, and often another is a GC-rich segment farther upstream. The general transcription factors that bind to the promoter-proximal elements are expressed in most cells, and sothey are available to initiate transcription at any time. Mutations in these sites can have a dramatic effect on transcription, demonstrating how important they are.

To modulate transcription, regulatory proteins possess one or more of the following functional domains:

1. A domain that recognizes a DNA regulatory sequence (the protein's DNA-binding site)
2. A domain that interacts with one or more proteins of the transcriptional apparatus (RNA polymerase or a protein associated with RNA polymerase)
3. A domain that interacts with proteins bound to nearby regulatory sequences on DNA such that they can act cooperatively to regulate transcription
4. A domain that influences chromatin condensation either directly or indirectly
5. A domain that acts as a sensor of physiological conditions within the cell Much of the strategy of eukaryotic transcriptional control hinges on how specific transcription factors control the access of general transcription factors and RNA polymerase II. Eukaryotic gene regulatory mechanisms have been discovered through both biochemical and genetic approaches. The latter has been advanced in particular by studies of the single-celled yeast *Saccharomyces cerevisiae*. Several decades of research have been a source of many insights into general principles of how eukaryotic transcriptional regulatory proteins work and how different cell types are generated. We'll examine two yeast gene regulatory systems in detail: the first concerns the galactose-utilization pathway; the second is the control of mating type.

Model Organism *Yeast*

Saccharomyces cerevisiae, or budding yeast, has emerged in recent years as the premier eukaryotic genetic system. Humans have grown yeast for centuries because it is an essential component of beer, bread, and wine. Yeast has many features that make it an ideal model organism. As a unicellular eukaryote, it can be grown on agar plates and, with a life cycle of just 90 minutes, large quantities can be cultured in liquid media. It has a very compact genome with only about 12 megabase pairs of DNA (compared with almost 3000 megabase pairs for humans) containing approximately 6000 genes that are distributed on 16 chromosomes. It was the first eukaryote to have its genome sequenced.

The yeast life cycle makes it very versatile for laboratory studies. Cells can be grown as either diploid or haploid. In both cases, the mother cell produces a bud containing an identical daughter cell. Diploid cells either continue to grow by budding or are induced to undergo meiosis, which produces four haploid spores held together in an *ascus* (also called a *tetrad*). Haploid spores of opposite mating type (a or ‘) will fuse and form a diploid. Spores of the same mating type will continue growth by budding.

Yeast has been called the *E. coli* of eukaryotes because of the ease of forward and reverse mutant analysis. To isolate mutants by using a forward genetic approach, haploid cells are mutagenized (with X rays, for example) and screened on plates for mutant phenotypes. This procedure is usually done by first plating cells on a rich medium on which all cells grow and by copying, or *replica plating,* the colonies from this master plate onto replica plates containing selective media or special growth conditions. For example, temperature-sensitive mutants will grow on the master plate at the permissive temperature but not on a replica plate at a restrictive temperature.

Comparison of the colonies on the master and replica plates will reveal the temperature-sensitive mutants. Using reverse genetics, scientists can also replace any yeast gene (of known or unknown function) with a mutant version (synthesized in a test tube) to understand the nature of the gene product.

Lessons from Yeast: The GAL System

To make use of extracellular galactose, yeast imports the sugar and converts it into a form of glucose that can be metabolized. Several genes—*GAL1, GAL2, GAL7,* and *GAL10*—in the yeast genome encode enzymes that catalyze steps in the biochemical pathway that converts galactose into glucose.

Three additional genes—*GAL3, GAL4,* and *GAL80*—encode proteins that regulate the expression of the enzyme genes. Just as in the *lac* system, the abundance of the sugar determines the level of gene expression in the biochemical pathway. In yeast cells growing in media lacking galactose, the *GAL* genes are largely silent. But, in the presence of galactose (and the absence of glucose), the *GAL* genes are induced. Just as for the *lac* operon, genetic and molecular analyses of mutants have been key to understanding how the expression of the genes in the galactose pathway is controlled.

The key regulator of *GAL* gene expression is the Gal4 protein, a sequencespecific DNA-binding protein. Gal4 is perhaps the best-studied transcriptional regulatory protein in eukaryotes. The detailed dissection of its regulation and activity has been a source of several key insights into the control of transcription in eukaryotes.

Gal4 Regulates Multiple Genes through Upstream Activating Sequences

In the presence of galactose, the *GAL1, GAL2, GAL7,* and *GAL10* genes are induced 1000-fold or more. In *GAL4* mutants, however, they remain silent. Each of these four genes has two or more Gal4-binding sites located 5_ (upstream) of its promoter. Consider the *GAL10* and *GAL1* genes, which are adjacent to each other and transcribed in opposite directions. Between the *GAL1* transcription start site and the *GAL10* transcription start site is a single 118-bp region that contains four Gal4- binding sites. Each Gal4-binding site is 17 base pairs long and is bound by one Gal4 protein dimer.

There are two Gal4-binding sites upstream of the *GAL2* gene as well, and another two upstream of the *GAL7* gene. These binding sites are required for gene activation in vivo. If they are deleted, the genes are silent, even in the presence of galactose. These regulatory sequences are enhancers that are also referred to as upstream activating sequences. The presence of enhancers located at a considerable linear distance from a eukaryotic gene's promoter is typical.

The level of gene expression in the biochemical pathway. In yeast cells growing in media lacking galactose, the *GAL* genes are largely silent. But, in the presence of galactose (and the absence of glucose), the *GAL* genes are induced. Just as for the *lac* operon, genetic and molecular analyses of mutants have been key to understanding how the expression of the genes in the galactose pathway is controlled.

The key regulator of *GAL* gene expression is the Gal4 protein, a sequencespecific DNA-binding protein. Gal4 is perhaps the best-studied transcriptional regulatory protein in eukaryotes. The detailed dissection of its regulation and activity has been a source of several key insights into the control of transcription in eukaryotes.

The Gal4 Protein has Separable DNA-binding and Activation Domains

After Gal4 is bound to the UAS element, how is gene expression induced? A distinct domain of the Gal4 protein, the activation domain, is required for regulatory activity. Thus, the Gal4 protein has at least two domains: one for DNA binding and another for activating transcription. A similar modular organization has been found to be a common feature of other DNA-binding transcription factors as well.

The modular organization of the Gal4 protein was demonstrated in a series of simple, elegant experiments. The strategy was to test the DNA binding and gene activation of mutant forms of the protein in which parts had been either deleted or fused to other proteins. In this manner, whether a part of the protein was necessary for a particular function could be determined. To carry out these studies, experimenters needed a simple means of assaying the expression of the enzymes encoded by the *GAL* genes.

The expression of *GAL* genes and other targets of transcription factors is typically monitored by using a reporter gene whose expression is easily tracked. Often, the reporter gene is the *lacZ* gene of *E. coli,* which can act on substrates whose products are easily measured by their bright color or fluourescence. Another common reporter gene is the gene that encodes the green fluorescent protein of jellyfish, which, as its name suggests, is easily tracked by the light that it emits. The coding region of one of these reporter genes and a promoter are placed downstream of a UAS element from a *GAL* gene.

Reporter expression is then a read-out of Gal4 activity in cells. When a form of the Gal4 protein lacking the activation domain is expressed in yeast, the binding sites of the UAS element are occupied, but no transcription is stimulated. The same is true when other regulatory proteins lacking activation domains, such as the bacterial repressor LexA, are expressed in cells bearing reporter genes with their respective binding sites. The more interesting result is obtained when a form of the Gal4 protein lacking the DNA-binding domain is grafted to the LexA DNA-binding domain; the hybrid protein now activates transcription from LexAbinding sites. Further "domain swap" experiments have revealed that the transcriptional activation function of the Gal4 protein resides in two small domains about 50 to 100 amino acids in length. These domains are separable from those used in the dimerization of the protein, DNA binding, and interaction with the Gal80 protein. The activation domain helps recruit the transcriptional machinery to the promoter.

This highly modular arrangement of activityregulating domains is found in many transcription factors.

Gal4 Activity is Physiologically Regulated

How does Gal4 become active in the presence of galactose? Key clues came from analyses of mutations in the *GAL80* and *GAL3* genes. In *GAL80* mutants, the *GAL* structural genes are active even in the absence of galactose. This result suggests that the normal function of the Gal80 protein is to somehow inhibit *GAL* gene expression. Conversely, in *GAL3* mutants, the *GAL* structural genes are not active in the presence of galactose, suggesting that Gal3 normally promotes expression of the *GAL* genes.

Extensive biochemical analyses have revealed that the Gal80 protein binds to the Gal4 protein with high affinity and directly inhibits Gal4 activity. Specifically, Gal80 binds to a region within one of the Gal4 activation domains, blocking its ability to promote the transcription of target genes. The role of the Gal3 protein is to release Gal4 from its inhibition by Gal80 in the presence of galactose. Gal3 is a sensor and inducer. When Gal3 binds galactose and ATP, it undergoes an allosteric change that promotes binding to Gal80, which in turn causes Gal80 to release Gal4, which is then able to activate transcription of its target genes. Thus, Gal3, Gal80, and Gal4 are all part of a switch whose state is determined by the presence or absence of galactose. In this switch, DNA binding by the transcriptional regulator is not the physiologically regulated step (as is the case in the *lac* operon and bacteriophage); rather, the activity of the activation domain is regulated.

Gal4 Functions in Most Eukaryotes

In addition to its action in yeast cells, Gal4 has been shown to be able to activate transcription in insect cells, human cells, and many other eukaryotic species. This versatility suggests that biochemical machinery and mechanisms of gene activation are common to a broad array of eukaryotes and that features revealed in yeast are generally present in other eukaryotes, and vice versa. Furthermore, because of their versatility, Gal4 and its UAS elements have become favored tools in genetic analysis for manipulating gene expression and function in a wide variety of model systems.

Activators Recruit the Transcriptional Machinery

In bacteria, activators commonly stimulate transcription by interacting directly with DNA and with RNA polymerase. In eukaryotes, activators generally work indirectly to recruit RNA polymerase II to gene promoters through two major mechanisms. First, activators can interact with subunits of the protein complexes having roles in transcription initiation. Second, activators can recruit proteins that modify chromatin structure, allowing RNA polymerase II and other proteins access to the DNA. Many activators, including Gal4, have both activities.

The eukaryotic transcriptional machinery contains many proteins that are parts of various subcomplexes within the transcriptional apparatus that is assembled on gene promoters. One subcomplex, transcription factor IID (TFIID), binds to the TATA box of eukaryotic promoters through the TATAbinding protein. Gal4 binds to TBP at a site in its activation domain, and, through this binding, it recruits the TFIID complex and, in turn, RNA polymerase II to the promoter.

The affinity of this interaction correlates well with Gal4's potency as an activator. Gal4 also interacts with the large Mediator complex, which directly interacts with RNA polymerase II to recruit it to gene promoters. The Mediator complex is an example of a coactivator, a term applied to a protein or protein complex that facilitates gene activation by a transcription factor but that itself is neither part of the transcriptional machinery nor a DNA-binding protein.

The ability of activators to bind to upstream DNA sequences and to interact with proteins that bind directly or indirectly to promoters helps to explain how transcription can be stimulated from more distant regulatory sequences.

Dynamic Chromatin and Eukaryotic Gene Regulation

A second mechanism for influencing gene transcription in eukaryotes modifies the local chromatin structure around gene regulatory sequences. To fully understand how this mechanism works, we need to first review chromatin structure and then consider how it can change and how these changes affect gene expression. The recruitment of transcriptional machinery by activators may appear to be somewhat similar in eukaryotes and bacteria, with the major difference being in the number of interacting proteins in the transcriptional machinery. Indeed, less than a decade ago, many biologists pictured eukaryotic regulation simply as a biochemically more complicated version of what had been discovered in bacteria. However, this view has changed dramatically as biologists have considered the effect of the organization of genomic DNA in eukaryotes.

Compared with eukaryotic DNA, bacterial DNA is relatively "naked," making it readily accessible to RNA polymerase. In contrast, eukaryotic chromosomes are packaged into chromatin, which is composed of DNA and proteins (mostly histones).

The basic unit of chromatin is the nucleosome, containing about 150 bp of DNA wrapped twice around a histone octamer. The histone octamer is composed of two subunits of each of the four histones: histone 2A, 2B, 3, and 4. Nucleosomes can associate into higherorder structures that further condense the DNA. The packaging of eukaryotic DNA into chromatin means that much of the DNA is not readily accessible to regulatory proteins and the transcriptional apparatus. Thus, whereas prokaryotic genes are generally accessible and "on" unless repressed, eukaryotic genes are inaccessible and "off" unless activated. Therefore, the modification of chromatin structure is a distinctive feature of eukaryotic gene regulation.

One can imagine several ways to alter chromatin structure. For example, one mechanism might be to simply move the histone octamer along the DNA. In the 1980s, biochemical techniques were developed that allowed researchers to determine the position of nucleosomes in and around specific genes. In these studies, chromatin was isolated from tissues or cells in which a gene was on and compared with chromatin from tissue where the same gene was off. The result for most genes analyzed was that nucleosome positions changed, especially in a gene's regulatory regions.

Thus, which DNA regions are wrapped up in nucleosomes can change: Nucleosome positions can shift on the DNA from cell to cell and over the life cycle of an organism. Transcription might be repressed when the promoter and flanking sequences are wound up in a nucleosome and inaccessible to RNA polymerase II. Activation of transcription would thus require the blocking nucleosome to be reorganized by nudging the histones or removing them entirely. Conversely, when gene repression is necessary, histone octamers may shift into a position that prevents transcription. The changing of nucleosome position is referred to as chromatin remodeling. Now, chromatin remodeling is known to be an integral part of eukaryotic gene expression, and great advances are being made in determining the underlying mechanism(s) and the regulatory proteins taking part. Here, again, genetic studies in yeast have been pivotal.

Chromatin-remodeling Proteins and Gene Activation

What was the connection between the ability to utilize sugar and the ability to switch mating types? The Snf2–Swi2 protein was purified and discovered to be part of a large, multisubunit complex called the SWI–SNF complex that can reposition nucleosomes in a test-tube assay if ATP is provided as an energy source. In some situations, the multisubunit SWI–SNF complex activates transcription by moving nucleosomes that are covering the TATA sequences and, in this way, facilitates the binding of RNA polymerase II. The SWI–SNF complex is thus a coactivator.

Gal4 also binds to the SWI–SNF complex and recruits the chromatin-remodeling complex to activated promoters. Yeast strains containing a defective SWI–SNF complex show a reduced level of Gal4 activity. Why might an activator use multiple activation mechanisms? There are at least two reasons understood at present. The first is that the accessibility of target promoters may change at different stages of the cell cycle or in different cell types (in multicellular eukaryotes). For example, during mitosis, when chromatin is more condensed, genes are less accessible. At that stage, Gal4 must recruit the chromatin-remodeling complexes, whereas, at other times, such recruitment might not be required to activate gene expression.

A second reason is that many transcription factors act in combinations to control gene expression synergistically. We will see shortly that this combinatorial synergy is a result of the fact that chromatin-remodeling complexes and the transcriptional machinery are recruited more efficiently when multiple transcription factors act together.

Histones and Chromatin Remodeling

A histone code As already stated, most nucleosomes are composed of an octamer made up of two copies each of the four core histones. Histones are known to be the most conserved proteins in nature; that is, histones are almost identical in all eukaryotic organisms from yeast to plants to animals. This conservation contributed to the view that histones could not take part in anything more complicated than the packaging of DNA to fit in the nucleus. However, recall that DNA with its four bases also was considered too "dumb" a molecule to carry the blueprint for all organisms on Earth.

Of note is that the histone proteins are organized into the core octamer with their amino-terminal ends protruding from the nucleosome. These protruding ends are called histone tails. Since the early 1960s, specific lysine residues in the histone tails have been known to be able to be covalently modified by the attachment of acetyl and methyl groups.

These reactions take place after the histone protein has been translated and even after the histone has been incorporated into a nucleosome.

There are now known to be at least 150 different histone modifications that require a wide variety of molecules in addition to the acetyl and methyl groups already mentioned (for example, phosphorylation and ubiquitylation).

Histone acetylation, deacetylation, and gene expression The acetylation reaction is the best-characterized histone modification:

Note that the reaction is reversible, which means that acetyl groups can be added and removed from the same histone residue. With 44 histone lysine residues available to accept acetyl groups, the presence or absence of these groups can carry a tremendous amount of information. For this reason, the covalent modification of histone tails is said to be a histone code. Scientists coined the expression histone code because the covalent modification of histone tails is reminiscent of the genetic code. For the histone code, information is stored in the patterns of histone modification rather than in the sequence of nucleotides. With more than 150 known histone modifications, there are a huge number of possible patterns and their effects on chromatin structure and transcriptional regulation are just beginning to be deciphered. To add to this complexity, the code is likely not interpreted in precisely the same way in all organisms. For now, let's see how the acetylation of histone amino acids influences chromatin structure and gene expression.

Evidence had been accumulating for years that the histones associated with the nucleosomes of active genes are rich in acetyl groups (said to be hyperacetylated), whereas inactive genes are underacetylated (hypoacetylated). The enzyme responsible for adding acetyl groups, histone acetyltransferase (HAT), proved very difficult to isolate. When it was finally isolated and its protein sequence deduced, it was found to be an *ortholog* of a yeast transcriptional activator called GCN5 (meaning that it was encoded by the same gene in a different organism). Thus, the conclusion was that GCN5 is a histone acetyltransferase. It binds to the DNA in the regulatory regions of some genes and activates transcription by acetylating nearby histones. Various protein complexes that are recruited by transcriptional activators are now understood to possess a HAT activity.

How does histone acetylation facilitate changes in gene expression? There appear to be at least two mechanisms for doing so. First, the addition of acetyl groups to specific histone residues can alter the interaction in a nucleosome between the DNA and a histone octamer so that the octamer is more likely to slide along the DNA to a new position. Second, histone acetylation, in conjunction with other histone modifications, influences the binding of regulatory proteins to the DNA. The bound regulatory protein may take part in one of several functions that either directly or indirectly increase the frequency of transcription initiation.

Like other histone modifications, acetylation is reversible, and histone deacetylases (HDAT's) also have been identified.

Such proteins play key roles in gene repression. For example, in the presence of galactose and glucose, the activation of *GAL* genes is prevented by the Mig1 protein. Mig1 is a sequence-specific DNA-binding repressor that binds to a site between the UAS element and the promoter of the *GAL1* gene. Mig1 recruits a protein complex called Tup1 that contains a histone deacetylase and that represses gene transcription. The Tup1 complex is an example of a corepressor, which faciliates gene repression but is not itself a DNA-binding repressor. The Tup1 complex is also recruited by other yeast repressors, such as MAT?2, and counterparts of this complex are found in all eukaryotes.

Mechanism of Enhancer Action

The development of a complex organism requires that transcription levels be regulated over a wide range. Think of a regulation mechanism as more like a rheostat than an on-or-off switch. In eukaryotes, transcription levels are made finely adjustable by the clustering of binding sites into enhancers. Several different transcription factors or several molecules of the same transcription factor may bind to adjacent sites. The binding of these factors to sites that are the correct distance apart leads to an amplified, or superadditive, effect on activating transcription. When an effect is greater than additive, it is said to be synergistic.

The binding of multiple regulatory proteins to the multiple binding sites in an enhancer can catalyze the formation of an enhanceosome, a large protein complex that acts synergistically to activate transcription. To better understand what an enhanceosome is and how it acts synergistically, let's look at a specific example.

The a-interferon Enhanceosome

The human -interferon gene, which encodes the antiviral protein interferon, is one of the best-characterized genes in eukaryotes. It is normally switched off but is activated to very high levels of transcription on viral infection. The key to the activation of this gene is the assembly of transcription factors into an enhanceosome about 100 bp upstream of the TATA box and transcription start site. The regulatory proteins of the -interferon enhanceosome all bind to the same face of the DNA double helix. Binding to the other side of the helix are several architectural proteins that bend the DNA and allow the different regulatory proteins to touch one another and form an activated complex. When all of the regulatory proteins are bound and interacting correctly, they form a "landing pad," a highaffinity binding site for the protein CBP, a coactivator protein that also recruits the transcriptional machinery. The large CBP protein also contains an intrinsic histone acetylase activity that modifies nucleosomes and facilitates high levels of transcription.

Cooperative interactions help to explain several perplexing observations about enhancers. For example, they explain why mutating any one transcription factor or binding site dramatically reduces enhancer activity. They also explain why the distance between binding sites within the enhancer is such a critical feature. Furthermore, enhancers do not have to be close to the start site of transcription.

The Control of Yeast Mating Type: Combinatorial Interactions

Thus far, we have focused in this chapter on the regulation of single genes or a few genes in one pathway. In multicellular organisms, distinct cell types differ in the expression of hundreds of genes. The expression or repression of sets of genes must therefore be coordinated in the making of particular cell types. One of the best-understood examples of cell-type regulation in eukaryotes is the regulation of mating type in yeast. This regulatory system has been dissected by an elegant combination of genetics, molecular biology, and biochemistry. Mating type serves as an excellent model for understanding the logic of gene regulation in multicellular animals.

The yeast *Saccharomyces cerevisiae* can exist in any of three different cell types known as a, and 'a/'.

The two cell types a and ' are haploid and contain only one copy of each chromosome. Although the two haploid cell types cannot be distinguished by their appearance in the microscope, they can be differentiated by a number of specific cellular characteristics, principally their mating type. An cell mates only with an a cell and secretes an oligopeptide pheromone, or sex hormone, called factor that arrests a cells in the cell cycle. A cell of the a type mates only with an cell and secretes a pheromone, called a factor, that arrests cells.

The diploid a cell does not mate, is larger than the and a cells, and does not respond to the mating hormones. Genetic analysis of mutants defective in mating has shown that cell type is controlled by a single genetic locus, the mating-type locus, *MAT*. There are two alleles of the *MAT* locus: haploid a cells have the *MATa* allele, haploid ' cells have the *MATa* allele, and the a/ ' diploid has both alleles. Although mating type is under genetic control, certain strains switch their mating type, sometimes as frequently as every cell division.

Enhancer-blocking Insulators

A regulatory element, such as an enhancer, that can act over tens of thousands of base pairs could interfere with the regulation of nearby genes. To prevent such promiscuous activation, regulatory elements called enhancer-blocking insulators have evolved. When positioned between an enhancer and a promoter, enhancerblocking insulators prevent the enhancer from activating transcription at that promoter. Such insulators have no effect on the activation of other promoters that are not separated from their enhancers by the insulator. Several models have been proposed to explain how an insulator could block enhancer activity only when placed between an enhancer and a promoter.

Genomic Imprinting

The phenomenon of genomic imprinting was discovered almost 20 years ago in mammals. In genomic imprinting, certain autosomal genes have unusual inheritance patterns. For example, an *igf2* allele is expressed in a mouse only if it is inherited from the mouse's father—an example of maternal imprinting because a copy of the gene derived from the mother is inactive. Conversely, a mouse *H19* allele is expressed only if it is inherited from the mother; *H19* is an example of paternal imprinting because the paternal copy is inactive. The consequence of parental imprinting is that imprinted genes are expressed as if there were only one copy of the gene present in the cell even though there are two. Hence, imprinting is an example of monoallelic inheritance. Importantly, no changes are observed in the DNA sequences of imprinted genes; that is, the identical gene can be active or inactive in the progeny, depending on whether it was inherited from mom or dad.

If the DNA sequence of the gene does not correlate with activity, what does? The answer is that, that the DNA in the regulatory regions of imprinted genes is methylated in a sex-specific manner in the development of gametes. DNA methylation usually results from the enzymatic addition of methyl groups to the carbon-5 position of a specific cytosine residue.

Both DNA methylation marks and histone modification marks can be stably inheritable from one cell generation to the next. We will see later in the chapter how such marks are thought to be duplicated in the course of DNA replication. For now, suffice it to say that such heritable alteration, in which the DNA sequence itself is unchanged, is called epigenetic inheritance, and the alterations (including both DNA methylation and histone modifications) are called epigenetic marks.

Let's turn again to the mouse *ifg2* and *H19* genes to see how imprinting works at the molecular level. These two genes are located in a cluster of imprinted genes on mouse chromosome 7. There are an estimated 100 imprinted genes in the mouse, and most are found in clusters comprising from 3 to 11 imprinted genes. (Humans have most of the same clustered imprintedgenes as those in the mouse.) In all cases examined, there is a specific DNA methylation pattern for each gene copy of an imprinted gene. For the *ifg2–H19* cluster, a specific region of DNA lying between the two genes is methylated in male germ cells and unmethylated in female germ cells. This region is called the imprinting control region (ICR). Only the unmethylated (female) ICR can bind a regulatory protein called CTCF. When bound, CTCF acts as an enhancer-blocking insulator that prevents enhancer activation of *Igf2* transcription. However, the enhancer in females can still activate *H19* transcription. In males, CTCF cannot bind to the ICR and the enhancer can activate *Igf2* transcription (recall that enhancers can act at great distances). The enhancer cannot activate *H19,* however, because the methylated region extends into the *H19* promoter. The methylated promoter cannot bind proteins needed for the transcription of *H19.*

Thus, we see how an enhancer-blocking insulator (in this case, CTCF bound to part of the ICR) prevents the enhancer from activating a distant gene (in this case, *Igf2*). Furthermore, we see that the CTCF-binding site is methylated only in chromosomes derived from the male parent. The methylation of the CTCF-binding site prevents CTCF binding in males and permits the enhancer to activate *Igf2.* Note that parental imprinting can greatly affect pedigree analysis. Because the inherited allele from one parent is inactive, a mutation in the allele inherited from the other parent will appear to be dominant, whereas, in fact, the allele is expressed because only one of the two homologs is active for this gene.

Many steps are required for imprinting. Soon after fertilization, mammals set aside cells that will become their germ cells. Imprints are removed or erased before the germ cells form. Without their distinguishing mark of DNA methylation, these genes are now said to be *epigenetically equivalent.* As these primordial germ cells become fully formed gametes, imprinted genes receive the sex-specific mark that will determine whether the gene will be active or silent after fertilization.

But what about Dolly and other Cloned Mammals?

Genomic imprinting leads to what many thought would be a requirement for the participation of male and female germ cells in mammalian embryo development. That is, male and female gametes contain different subsets of imprinted genes so that the embryo will have a full complement of active imprinted genes. Why then are mammals such as Dolly and, more recently, cloned pigs, cats, dogs, and cows that were derived from somatic nuclei able to survive and even flourish? After all, as already noted, the mutation of even a single imprinted gene can be lethal or can lead to serious disease.

At this point, scientists do not understand why the cloning of many mammalian species has been successful. However, despite these successes, cloning is extremely inefficient in all species tested. For most experiments, a successful clone is an exceedingly rare event, requiring hundreds, even thousands, of attempts. One could argue that the failure of most cloned embryos to develop into viable organisms is a testament to the importance of the epigenetic mechanisms of gene regulation in eukaryotes. As such, it illustrates how knowledge of the complete DNA sequence of all genes in an organism is only a first step in understanding how eukaryotic genes are regulated.

Chromatin Domains and Their Inheritance

Thus far, we have looked at how genes are activated in a chromatin environment. However, as stated at the beginning of this chapter, most of the genes in eukaryotic genomes are off at any one time. Let's now turn to those vast regions of the genome that are transcriptionally inactive. One of the most useful models for understanding mechanisms that maintain the inactivity of genes concerns the control of yeast mating type and mating-type switching.

Mating-type Switching and Gene Silencing

Haploid yeast cells are able to switch their mating type. Genetic analyses of certain mutants that either could not switch or could not mate (they were sterile) were sources of key insights into mating-type switching. Among the switch mutants were several mutant loci including the *HO* gene and the *HMRa* and *HMLa* genes. Further study revealed that the *HO* gene encodes an endonuclease, an enzyme that cleaves DNA, required for the initiation of switching. It was also found that the *HMRa* and *HMLa* loci contain "cassettes" of unexpressed genetic information for the MATa and MAT??mating types, respectively. The *HMR* and *HML* loci are thus referred to as "silent" cassettes.

The HO endonuclease initiates the mating-type switch by inserting a doublestrand break at the *MAT* locus. The interconversion of mating type then takes place by a type of recombination between the segment of DNA (a cassette) from one of the two unexpressed loci and the *MAT* locus. The result is the replacement of the old cassette at the *MAT* locus with a new cassette. The resulting mating type is either the MATa or the MAT??type, depending on which cassette is at the *MAT* locus. The inserted cassette is actually copied from the *HML* or *HMR* locus.

In this manner, the switch is reversible because the information for the *a* and *a* cassettes is always present at the *HMR* and *HML* loci and never lost. Normally, the *HMR* and *HML* cassettes are "silent." However, in SIR mutants (silent information regulators), silencing is compromised such that both *a* and *a* information is expressed. The resulting mutants are sterile. The Sir2, Sir3, and Sir4 proteins form a complex that plays a key role in gene silencing. Sir2 is a histone deacetylase that facilitates the condensation of chromatin and helps lock up *HMR* and *HML* in chromatin domains that are inaccessible to transcriptional activators. Gene silencing is a very different process from gene repression; silencing is a position effect that depends on the neighborhood in which genetic information is located. You will learn more about position effects later, in the section on positioneffect variegation in the fruit fly *Drosophila melanogaster*.

Heterochromatin and Euchromatin Compared

Let's return to the silent cassettes *HML* and *HMR* to understand why gene silencing is a very different process from gene repression and what is meant by a genomic neighborhood. To do so, it is important to note that chromatin is not uniform over all chromosomes; certain domains of chromosomes are bundled in highly condensed chromatin called heterochromatin. Other domains are packaged in lesscondensed chromatin called euchromatin. Chromatin condensation also changes in the course of the cell cycle.

The chromatin of cells entering mitosis becomes highly condensed as the chromosomes align in preparation for cell division. After cell division, regions forming heterochromatin remain condensed especially around the centromeres and telomeres (called constitutive heterochromatin), whereas the regions forming euchromatin become less condensed. Geneticists first suspected a limited role for the influence of chromatin structure on gene regulation early in the history of genetics. At that time, they noticed that heterochromatic DNA contained few genes, whereas euchromatin was rich in genes. But what is heterochromatin if not genes? Most of the eukaryotic genome is composed of repetitive sequences that do not make protein or structural RNA—sometimes called junk DNA.

Thus, the densely packed nucleosomes of heterochromatin were said to form a "closed" structure that was inaccessible to regulatory proteins and inhospitable to gene activity. In contrast, euchromatin, with its more widely spaced nucleosomes, was proposed to assume an "open" structure that permitted transcription. The existence of open and closed regions of chromatin was also suggested as a reason that recombination frequencies are 100- to 1000-fold higher in euchromatin compared with heterochromatin. Euchromatin, with its more open conformation, was hypothesized to be more accessible to proteins needed for DNA recombination.

Position-effect Variegation in *Drosophila* Reveals Genomic Neighborhoods

The geneticist Hermann Muller first discovered an interesting genetic phenomenon while studying *Drosophila:* there exist chromosomal neighborhoods that can silence genes that are experimentally "relocated" to adjacent regions of the chromosome.

In these experiments, flies were irradiated with X rays to induce mutations in their germ cells. The progeny of the irradiated flies were screened for unusual phenotypes. A mutation in the *white* gene, near the tip of the X chromosome, will result in progeny with white eyes instead of the wild-type red color. Some of the progeny had very unusual eyes with patches of white and red color. Cytological examination revealed a chromosomal rearrangement in the mutant flies: present in the X chromosome was an inversion of a piece of the chromosome carrying the *white* gene. In this rearrangement, the *white* gene, which is normally located in a euchromatic region of the X chromosome, now finds itself near the heterochromatic centromere. In some cells, the heterochromatin can "spread" to the neighbouring euchromatin and silences the *white* gene. Patches of white tissue in the eye are derived from the descendants of a single cell in which the *white* gene has been epigenetically silenced and remains silenced through future cell divisions.

In contrast, the red patches arise from cells in which heterochromatin has not spread to the *white* gene, and so this gene remains active in all its descendants. The existence of red and white patches of cells in the eye of a single organism dramatically illustrates two features of epigenetic silencing. First, that the expression of a gene can be repressed by virtue of its position in the chromosome rather than by a mutation in its DNA sequence. Second, that epigenetic silencing can be inherited from one cell generation to the next. Findings from subsequent studies in *Drosophila* and yeast demonstrated that many active genes are silenced in this mosaic fashion when they are relocated to neighborhoods (near centromeres or telomeres) that are heterochromatic.

Thus, the ability of heterochromatin to spread into euchromatin and silence genes is a feature common to many organisms. This phenomenon has been called positioneffect variegation (PEV). It provides powerful evidence that chromatin structure is able to regulate the expression of genes—in this case, determining whether genes with identical DNA sequence will be active or silenced.

Genetic Analysis of PEV Reveals Proteins Necessary for Heterochromatin Formation

To find out what proteins might be implicated in the establishment of heterochromatin, geneticists isolated mutations at a second chromosomal locus that either suppressed or enhanced the variegated pattern. Suppressors of variegation [called *Su(var)*] are genes that, when mutated, reduce the spread of heterochromatin, meaning that the wild-type products of these genes are required for spreading. In fact, the *Su(var)* alleles have proved to be a treasure trove for scientists interested in the proteins that are required to establish and maintain the inactive, heterochromatic state.

Among more than 50 *Drosophila* gene products identified by these screens was heterochromatin protein-1 (HP-1), which had previously been found associated with the heterochromatic telomeres and centromeres. Thus, it makes sense that a mutation in the gene encoding HP-1 will show up as a *Su(var)* allele because the protein is required in some way to produce or maintain heterochromatin. Another *Su(var)* gene was found to encode a methyltransferase that adds methyl groups to a specific amino acid residue in the tail of histone H3 (called histone H3 methyltransferase or HMTase). Proteins similar to HP-1 and HMTase have been isolated in diverse taxa, suggesting the conservation of an important eukaryotic function.

Silencing an Entire Chromosome: X-chromosome Inactivation

The epigenetic phenomenon called X-chromosome inactivation has intrigued scientists for decades. For now, it is sufficient to know that the number of transcripts produced by a gene is usually proportional to the number of copies of that gene in a cell. Mammals, for example, are diploid and have two copies of each gene located on their autosomes. For the vast majority of genes, both alleles are expressed. However, this is not possible for the sex chromosomes.

The number of the X and Y sex chromosomes differs between the sexes, with female mammals having two X chromosomes and males having only one. The mammalian X chromosome is thought to contain about 1000 genes. Females have twice as many copies of these X-linked genes and would otherwise express twice as much transcript from these genes as males do if there were not a mechanism to correct this imbalance. (Not having a Y chromosome is not a problem for females, because the very few genes on this chromosome are required only for the development of males.) This dosage imbalance is corrected by a process called dosage compensation, which makes the amount of most gene products from the two copies of the X chromosome in females equivalent to the single dose of the X chromosome in males.

In mammals, this equivalency is accomplished by random inactivation of one of the two X chromosomes in each cell at an early stage in development. This inactive state is then propagated to all progeny cells. (In the germ line, the second X chromosome becomes reactivated in oogenesis). The inactivated chromosome, called a Barr body, can be seen in the nucleus as a darkly staining, highly condensed, heterochromatic structure. Two aspects of X-chromosome inactivation are relevant to a discussion of chromatin and the regulation of gene expression. First, most of the genes on the inactivated X chromosome are silenced, and the chromosome has epigenetic marks associated with heterochromatin including methylation of H3 at lysine 9 and hypermethylation of its DNA. Second, genes on the inactivated chromosome remain inactive in all descendants of these cells. Because the DNA sequence itself is unchanged, this heritable alteration is an example of epigenetic inheritance.

The Inheritance of Epigenetic Marks and Chromatin Structure

Epigenetic inheritance can be defined operationally as the inheritance of chromatin states from one cell generation to the next. What this inheritance means is that, in DNA replication, both the DNA sequence and the chromatin structure are faithfully passed on to the next cell generation. However, unlike the sequence of DNA, chromatin structure can change in the course of the cell cycle when, for example, transcription factors modify the histone code, causing local changes in nucleosome position or nucleosome density or both. The replisome not only copies the parental strands but also disassembles the nucleosomes in the parental strands and reassembles them in both the parental and the daughter strands. This process is accomplished by the random distribution of the old histones from existing nucleosomes to daughter strands and the delivery of new histones to the replisome. In this way, the old histones with their modified tails and the new histones with unmodified tails are assembled into nucleosomes that become associated with both daughter strands. The code carried by the old histones most likely guides the modification of the new histones.

The inheritance of DNA methylation is better understood. Semiconservative replication generates daughter helices that are methylated on one of their two strands (the parental strand). The unmethylated strands are methylated by DNA methyltransferases that have a high affinity for these so-called hemimethylated substrates and are guided by the methylation pattern on the parental strand. Thus, the information inherent in the histone code and the existing DNA methylation patterns serve to reconstitute the local chromatin structure that existed before DNA synthesis and mitosis.

Mutations

Mutations are changes in a genomic sequence: the DNA sequence of a cell's genome or the DNA or RNA sequence of a virus. Mutations are caused by radiation, viruses, transposons and mutagenic chemicals, as well as errors that occur during meiosis or DNA replication. They can also be induced by the organism itself, by cellular processes such as hypermutation.

Mutation can result in several different types of change in DNA sequences; these can either have no effect, alter the product of a gene, or prevent the gene from functioning properly or completely. Studies in the fly *Drosophila melanogaster* suggest that if a mutation changes a protein produced by a gene, this will probably be harmful, with about 70 percent of these mutations having damaging effects, and the remainder being either neutral or weakly beneficial. Due to the damaging effects that mutations can have on cells, organisms have evolved mechanisms such as DNA repair to remove mutations. Therefore, the optimal mutation rate for a species is a trade-off between costs of a high mutation rate, such as deleterious mutations, and the metabolic costs of maintaining systems to reduce the mutation rate, such as DNA repair enzymes. Viruses that use RNA as their genetic material have rapid mutation rates, which can be an advantage since these viruses will evolve constantly and rapidly, and thus evade the defensive responses of e.g. the human immune system.

Description

Mutations can involve large sections of DNA becoming duplicated, usually through genetic recombination. These duplications are a major source of raw material for evolving new genes, with tens to hundreds of genes duplicated in animal genomes every million years. Most genes belong to larger families of genes of shared ancestry. Novel genes are produced by several methods, commonly through the duplication and mutation of an ancestral gene, or by recombining parts of different genes to form new combinations with new functions. Here, domains act as modules, each with a particular and independent function, that can be mixed together to produce genes encoding new proteins with novel properties.

For example, the human eye uses four genes to make structures that sense light: three for color vision and one for night vision; all four arose from a single ancestral gene. Another advantage of duplicating a gene (or even an entire genome) is that this increases redundancy; this allows one gene in the pair to acquire a new function while the other copy performs the original function. Other types of mutation occasionally create new genes from previously non-coding DNA.

Changes in chromosome number may involve even larger mutations, where segments of the DNA within chromosomes break and then rearrange. For example, two chromosomes in the *Homo* genus fused to produce human chromosome 2; this fusion did not occur in the lineage of the other apes, and they retain these separate chromosomes. In evolution, the most important role of such chromosomal rearrangements may be to accelerate the divergence of a population into new species by making populations less likely to interbreed, and thereby preserving genetic differences between these populations.

Sequences of DNA that can move about the genome, such as transposons, make up a major fraction of the genetic material of plants and animals, and may have been important in the evolution of genomes. For example, more than a million copies of the Alu sequence are present in the human genome, and these sequences have now been recruited to perform functions such as regulating gene expression. Another effect of these mobile DNA sequences is that when they move within a genome, they can mutate or delete existing genes and thereby produce genetic diversity.

In multicellular organisms with dedicated reproductive cells, mutations can be subdivided into germ line mutations, which can be passed on to descendants through their reproductive cells, and somatic mutations (also called acquired mutations), which involve cells outside the dedicated reproductive group and which are not usually transmitted to descendants. If the organism can reproduce asexually through mechanisms such as cuttings or budding the distinction can become blurred. For example, plants can sometimes transmit somatic mutations to their descendants asexually or sexually where flower buds develop in somatically mutated parts of plants. A new mutation that was not inherited from either parent is called a *de novo* mutation. The source of the mutation is unrelated to the consequence, although the consequences are related to which cells were mutated.

Nonlethal mutations accumulate within the gene pool and increase the amount of genetic variation. The abundance of some genetic changes within the gene pool can be reduced by natural selection, while other "more favourable" mutations may accumulate and result in adaptive evolutionary changes.

For example, a butterfly may produce offspring with new mutations. The majority of these mutations will have no effect; but one might change the color of one of the butterfly's offspring, making it harder (or easier) for predators to see. If this color change is advantageous, the chance of this butterfly surviving and producing its own offspring are a little better, and over time the number of butterflies with this mutation may form a larger percentage of the population.

Neutral mutations are defined as mutations whose effects do not influence the fitness of an individual. These can accumulate over time due to genetic drift. It is believed that the overwhelming majority of mutations have no significant effect on an organism's fitness. Also, DNA repair mechanisms are able to mend most changes before they become permanent mutations, and many organisms have mechanisms for eliminating otherwise permanently mutated somatic cells.

Mutation is generally accepted by biologists as the mechanism by which natural selection acts, generating advantageous new traits that survive and multiply in offspring as well as disadvantageous traits, in less fit offspring, that tend to die out.

Causes

Two classes of mutations are spontaneous mutations (molecular decay) and induced mutations caused by mutagens.

Spontaneous mutations on the molecular level can be caused by:

- Tautomerism – A base is changed by the repositioning of a hydrogen atom, altering the hydrogen bonding pattern of that base resulting in incorrect base pairing during replication.
- Depurination – Loss of a purine base (A or G) to form an apurinic site (AP site).

- Deamination – Hydrolysis changes a normal base to an atypical base containing a keto group in place of the original amine group. Examples include C → U and A '! HX (hypoxanthine), which can be corrected by DNA repair mechanisms; and 5MeC (5-methylcytosine) → T, which is less likely to be detected as a mutation because thymine is a normal DNA base.
- Slipped strand mispairing - Denaturation of the new strand from the template during replication, followed by renaturation in a different spot ("slipping"). This can lead to insertions or deletions.

Induced mutations on the molecular level can be caused by:

- Chemicals
  - o Hydroxylamine NH_2OH
  - o Base analogs (e.g. BrdU)
  - o Alkylating agents (e.g. *N*-ethyl-*N*-nitrosourea) These agents can mutate both replicating and non-replicating DNA. In contrast, a base analog can only mutate the DNA when the analog is incorporated in replicating the DNA. Each of these classes of chemical mutagens has certain effects that then lead to transitions, transversions, or deletions.
  - o Agents that form DNA adducts (e.g. ochratoxin A metabolites)
  - o DNA intercalating agents (e.g. ethidium bromide)
  - o DNA crosslinkers
  - o Oxidative damage
  - o Nitrous acid converts amine groups on A and C to diazo groups, altering their hydrogen bonding patterns which leads to incorrect base pairing during replication.
- Radiation

o Ultraviolet radiation (nonionizing radiation). Two nucleotide bases in DNA – cytosine and thymine – are most vulnerable to radiation that can change their properties. UV light can induce adjacent pyrimidine bases in a DNA strand to become covalently joined as a pyrimidine dimer. UV radiation, particularly longer-wave UVA, can also cause oxidative damage to DNA.

o Ionizing radiation

- Viral infections.

DNA has so-called hotspots, where mutations occur up to 100 times more frequently than the normal mutation rate. A hotspot can be at an unusual base, e.g., 5-methylcytosine.

Mutation rates also vary across species. Evolutionary biologists have theorized that higher mutation rates are beneficial in some situations, because they allow organisms to evolve and therefore adapt more quickly to their environments. For example, repeated exposure of bacteria to antibiotics, and selection of resistant mutants, can result in the selection of bacteria that have a much higher mutation rate than the original population (mutator strains).

Classification of Mutation Types

By Effect on Structure

The sequence of a gene can be altered in a number of ways. Gene mutations have varying effects on health depending on where they occur and whether they alter the function of essential proteins. Mutations in the structure of genes can be classified as:

- Small-scale mutations, such as those affecting a small gene in one or a few nucleotides, including:
  - o Point mutations, often caused by chemicals or malfunction of DNA replication, exchange a single nucleotide for another. These changes are classified as transitions or transversions. Most common is the transition that exchanges a purine for a purine (A ↔ G) or a pyrimidine for a pyrimidine, (C ↔ T). A transition can be caused by nitrous acid, base mis-pairing, or mutagenic base analogs such as 5-bromo-2-deoxyuridine (BrdU). Less common is a transversion, which exchanges a purine for a pyrimidine or a pyrimidine for a purine (C/T ↔ A/G). An example of a transversion is adenine (A) being converted into a cytosine (C). A point mutation can be reversed by another point mutation, in which the nucleotide is changed back to its original state (true reversion) or by second-site reversion (a complementary mutation elsewhere that results in regained gene functionality). Point mutations that occur within the protein coding region of a gene may be classified into three kinds, depending upon what the erroneous codon codes for:
    - * Silent mutations: which code for the same amino acid.
    - * Missense mutations: which code for a different amino acid.
    - * Nonsense mutations: which code for a stop and can truncate the protein.
  - o Insertions add one or more extra nucleotides into the DNA. They are usually caused by transposable elements, or errors during replication of repeating elements (e.g. AT repeats). Insertions in the coding region of a gene may alter splicing of the mRNA (splice site

mutation), or cause a shift in the reading frame (frameshift), both of which can significantly alter the gene product. Insertions can be reverted by excision of the transposable element.

- o Deletions remove one or more nucleotides from the DNA. Like insertions, these mutations can alter the reading frame of the gene. They are generally irreversible: though exactly the same sequence might theoretically be restored by an insertion, transposable elements able to revert a very short deletion (say 1–2 bases) in *any* location are either highly unlikely to exist or do not exist at all. Note that a deletion is not the exact opposite of an insertion: the former is quite random while the latter consists of a specific sequence inserting at locations that are not entirely random or even quite narrowly defined.

- Large-scale mutations in chromosomal structure, including:
  - o Amplifications (or gene duplications) leading to multiple copies of all chromosomal regions, increasing the dosage of the genes located within them.
  - o Deletions of large chromosomal regions, leading to loss of the genes within those regions.
  - o Mutations whose effect is to juxtapose previously separate pieces of DNA, potentially bringing together separate genes to form functionally distinct fusion genes (e.g. bcr-abl). These include:
    - * Chromosomal translocations: interchange of genetic parts from nonhomologous chromosomes.
    - * Interstitial deletions: an intra-chromosomal deletion that removes a segment of DNA from a single chromosome, thereby apposing previously distant genes. For example, cells isolated from a human astrocytoma, a type of brain tumour, were found to have a chromosomal deletion removing sequences between the "fused in glioblastoma" gene and the receptor tyrosine kinase "ros", producing a fusion protein (FIG-ROS). The abnormal FIG-ROS fusion protein has constitutively active kinase activity that causes oncogenic transformation (a transformation from normal cells to cancer cells).
    - * Chromosomal inversions: reversing the orientation of a chromosomal segment.
  - o Loss of heterozygosity: loss of one allele, either by a deletion or recombination event, in an organism that previously had two different alleles.

By Effect on Function

- Loss-of-function mutations are the result of gene product having less or no function. When the allele has a complete loss of function (null allele) it is often called an amorphic mutation. Phenotypes associated with such mutations are most often recessive. Exceptions are when the organism is haploid, or when the reduced dosage of a normal gene product is not enough for a normal phenotype (this is called haploinsufficiency).
- Gain-of-function mutations change the gene product such that it gains a new and abnormal function. These mutations usually have dominant phenotypes. Often called a neomorphic mutation.
- Dominant negative mutations (also called antimorphic mutations) have an altered gene product that acts antagonistically to the wild-type allele. These mutations usually result in an altered molecular function (often inactive) and are characterised by a dominant or semi-dominant phenotype. In humans, Marfan syndrome is an example of a dominant negative mutation occurring in an autosomal dominant disease. In this condition, the defective glycoprotein product of the fibrillin gene (FBN1) antagonizes the product of the normal allele.
- Lethal mutations are mutations that lead to the death of the organisms which carry the mutations.
- A back mutation or reversion is a point mutation that restores the original sequence and hence the original phenotype.

By Effect on Fitness

In applied genetics it is usual to speak of mutations as either harmful or beneficial.

- A harmful mutation is a mutation that decreases the fitness of the organism.
- A beneficial mutation is a mutation that increases fitness of the organism, or which promotes traits that are desirable.

In theoretical population genetics, it is more usual to speak of such mutations as deleterious or advantageous. In the neutral theory of molecular evolution, genetic drift is the basis for most variation at the molecular level.

- A neutral mutation has no harmful or beneficial effect on the organism. Such mutations occur at a steady rate, forming the basis for the molecular clock.
- A deleterious mutation has a negative effect on the phenotype, and thus decreases the fitness of the organism.

- An advantageous mutation has a positive effect on the phenotype, and thus increases the fitness of the organism.
- A nearly neutral mutation is a mutation that may be slightly deleterious or advantageous, although most nearly neutral mutations are slightly deleterious.

In reality, viewing the fitness effects of mutations in these discrete categories is an oversimplification. Attempts have been made to infer the distribution of fitness effects using mutagenesis experiments or theoretical models applied to molecular sequence data.

However, the current distribution is still uncertain, and some aspects of the distribution likely vary between species.

By Inheritance

- inheritable generic in pro-generic tissue or cells on path to be changed to gametes.
- non inheritable somatic (e.g., carcinogenic mutation)
- non inheritable post mortem aDNA mutation in decaying remains.

By pattern of inheritance The human genome contains two copies of each gene – a paternal and a maternal allele.

- A heterozygous mutation is a mutation of only one allele.
- A homozygous mutation is an identical mutation of both the paternal and maternal alleles.
- Compound heterozygous mutations or a genetic compound comprises two different mutations in the paternal and maternal alleles.
- A wildtype or homozygous non-mutated organism is one in which neither allele is mutated. (Just not a mutation).

By Impact on Protein Sequence

- A frameshift mutation is a mutation caused by insertion or deletion of a number of nucleotides that is not evenly divisible by three from a DNA sequence. Due to the triplet nature of gene expression by codons, the insertion or deletion can disrupt the reading frame, or the grouping of the codons, resulting in a completely different translation from the original. The earlier in the sequence the deletion or insertion occurs, the more altered the protein produced is.
- A nonsense mutation is a point mutation in a sequence of DNA that results in a premature stop codon, or a *nonsense codon* in the transcribed mRNA, and possibly a truncated, and often nonfunctional protein product.

- Missense mutations or *nonsynonymous mutations* are types of point mutations where a single nucleotide is changed to cause substitution of a different amino acid. This in turn can render the resulting protein nonfunctional. Such mutations are responsible for diseases such as Epidermolysis bullosa, sickle-cell disease, and SOD1 mediated ALS (Boillée 2006, p. 39).
- A neutral mutation is a mutation that occurs in an amino acid codon which results in the use of a different, but chemically similar, amino acid. The similarity between the two is enough that little or no change is often rendered in the protein. For example, a change from AAA to AGA will encode lysine, a chemically similar molecule to the intended arginine.
- Silent mutations are mutations that do not result in a change to the amino acid sequence of a protein. They may occur in a region that does not code for a protein, or they may occur within a codon in a manner that does not alter the final amino acid sequence. The phrase *silent mutation* is often used interchangeably with the phrase *synonymous mutation*; however, synonymous mutations are a subcategory of the former, occurring only within exons. The name silent could be a misnomer. For example, a silent mutation in the exon/intron border may lead to alternative splicing by changing the splice site, thereby leading to a changed protein.

Special Classes

- Conditional mutation is a mutation that has wild-type (or less severe) phenotype under certain "permissive" environmental conditions and a mutant phenotype under certain "restrictive" conditions. For example, a temperature-sensitive mutation can cause cell death at high temperature (restrictive condition), but might have no deleterious consequences at a lower temperature (permissive condition).

Nomenclature

Nomenclature of mutations specify the type of mutation and base or amino acid changes.

- Nucleotide substitution (e.g. 76A>T) - The number is the position of the nucleotide from the 5' end, the first letter represents the wild type nucleotide, and the second letter represents the nucleotide which replaced the wild type. In the given example, the adenine at the 76th position was replaced by a thymine.
  - If it becomes necessary to differentiate between mutations in genomic DNA, mitochondrial DNA, and RNA, a simple convention is used.

For example, if the 100th base of a nucleotide sequence mutated from G to C, then it would be written as g.100G>C if the mutation occurred in genomic DNA, m.100G>C if the mutation occurred in mitochondrial DNA, or r.100g>c if the mutation occurred in RNA. Note that for mutations in RNA, the nucleotide code is written in lower case.

- Amino acid substitution (e.g. D111E) – The first letter is the one letter code of the wild type amino acid, the number is the position of the amino acid from the N terminus, and the second letter is the one letter code of the amino acid present in the mutation. Nonsense mutations are represented with an X for the second amino acid (e.g. D111X).
- Amino acid deletion (e.g. ÄF508) – The Greek letter Ä (delta) indicates a deletion. The letter refers to the amino acid present in the wild type and the number is the position from the N terminus of the amino acid were it to be present as in the wild type.

Harmful Mutations

Changes in DNA caused by mutation can cause errors in protein sequence, creating partially or completely non-functional proteins. To function correctly, each cell depends on thousands of proteins to function in the right places at the right times. When a mutation alters a protein that plays a critical role in the body, a medical condition can result. A condition caused by mutations in one or more genes is called a genetic disorder. Some mutations alter a gene's DNA base sequence but do not change the function of the protein made by the gene. Studies of the fly *Drosophila melanogaster* suggest that if a mutation does change a protein, this will probably be harmful, with about 70 percent of these mutations having damaging effects, and the remainder being either neutral or weakly beneficial. However, studies in yeast have shown that only 7% of mutations that are not in genes are harmful.

If a mutation is present in a germ cell, it can give rise to offspring that carries the mutation in all of its cells. This is the case in hereditary diseases. On the other hand, a mutation may occur in a somatic cell of an organism. Such mutations will be present in all descendants of this cell within the same organism, and certain mutations can cause the cell to become malignant, and thus cause cancer.

Often, gene mutations that could cause a genetic disorder are repaired by the DNA repair system of the cell. Each cell has a number of pathways through which enzymes recognize and repair mistakes in DNA. Because DNA can be damaged or mutated in many ways, the process of DNA repair is an important way in which the body protects itself from disease.

Beneficial Mutations

Although most mutations that change protein sequences are neutral or harmful, some mutations have a positive effect on an organism. In this case, the mutation may enable the mutant organism to withstand particular environmental stresses better than wild-type organisms, or reproduce more quickly. In these cases a mutation will tend to become more common in a population through natural selection.

For example, a specific 32 base pair deletion in human CCR5 (CCR5-Ä32) confers HIV resistance to homozygotes and delays AIDS onset in heterozygotes. The CCR5 mutation is more common in those of European descent. One possible explanation of the etiology of the relatively high frequency of CCR5-Ä32 in the European population is that it conferred resistance to the bubonic plague in mid-14th century Europe. People with this mutation were more likely to survive infection; thus its frequency in the population increased. This theory could explain why this mutation is not found in southern Africa, where the bubonic plague never reached. A newer theory suggests that the selective pressure on the CCR5 Delta 32 mutation was caused by smallpox instead of the bubonic plague.

Another example, is Sickle cell disease which is a blood disorder in which the body produces an abnormal type of the oxygen-carrying substance hemoglobin in the red blood cells. One-third of all indigenous inhabitants of Sub-Saharan Africa carry the gene, because in areas where malaria is common, there is a survival value in carrying only a single sickle-cell gene (sickle cell trait). Those with only one of the two alleles of the sickle-cell disease are more resistant to malaria, since the infestation of the malaria plasmodium is halted by the sickling of the cells which it infests.

Prion Mutation

Prions are proteins and do not contain genetic material. However, prion replication has been shown to be subject to mutation and natural selection just like other forms of replication.

Cancer

Cancer (medical term: malignant neoplasm) is a class of diseases in which a group of cells display *uncontrolled growth* (division beyond the normal limits), *invasion* (intrusion on and destruction of adjacent tissues), and sometimes *metastasis* (spread to other locations in the body via lymph or blood). These three malignant properties of cancers differentiate them from benign tumors, which are self-limited, and do not invade or metastasize. Most cancers form a tumour but some, like leukemia, do not. The branch of medicine concerned with the study, diagnosis, treatment, and prevention of cancer is oncology. Cancer affects people at all ages with the risk for most types increasing with age. Cancer caused about 13% of all human deaths in 2007 (7.6 million).

Cancers are caused by abnormalities in the genetic material of the transformed cells. These abnormalities may be due to the effects of carcinogens, such as tobacco smoke, radiation, chemicals, or infectious agents. Other cancer-promoting genetic abnormalities may randomly occur through errors in DNA replication, or are inherited, and thus present in all cells from birth. The heritability of cancers is usually affected by complex interactions between carcinogens and the host's genome.

Genetic abnormalities found in cancer typically affect two general classes of genes. Cancer-promoting *oncogenes* are typically activated in cancer cells, giving those cells new properties, such as hyperactive growth and division, protection against programmed cell death, loss of respect for normal tissue boundaries, and the ability to become established in diverse tissue environments. *Tumour suppressor genes* are then inactivated in cancer cells, resulting in the loss of normal functions in those cells, such as accurate DNA replication, control over the cell cycle, orientation and adhesion within tissues, and interaction with protective cells of the immune system.

Definitive diagnosis requires the histologic examination of a biopsy specimen, although the initial indication of malignancy can be symptomatic or radiographic imaging abnormalities. Most cancers can be treated and some forced into remission, depending on the specific type, location, and stage. Once diagnosed, cancer is usually treated with a combination of surgery, chemotherapy and radiotherapy. As research develops, treatments are becoming more specific for different varieties of cancer. There has been significant progress in the development of targeted therapy drugs that act specifically on detectable molecular abnormalities in certain tumors, and which minimize damage to normal cells. The prognosis of cancer patients is most influenced by the type of cancer, as well as the stage, or extent of the disease. In addition, histologic grading and the presence of specific molecular markers can also be useful in establishing prognosis, as well as in determining individual treatments.

Classification

Cancers are classified by the type of cell that resembles the tumour and, therefore, the tissue presumed to be the origin of the tumour. These are the histology and the location, respectively. Examples of general categories include:

- Carcinoma: Malignant tumors derived from epithelial cells. This group represents the most common cancers, including the common forms of breast, prostate, lung and colon cancer.

- Sarcoma: Malignant tumors derived from connective tissue, or mesenchymal cells.
- Lymphoma and leukemia: Malignancies derived from hematopoietic (blood-forming) cells
- Germ cell tumour: Tumors derived from totipotent cells. In adults most often found in the testicle and ovary; in fetuses, babies, and young children most often found on the body midline, particularly at the tip of the tailbone; in horses most often found at the poll (base of the skull).
- Blastic tumour or blastoma: A tumour (usually malignant) which resembles an immature or embryonic tissue. Many of these tumors are most common in children.

Malignant tumors (cancers) are usually named using -carcinoma, -sarcoma or -blastoma as a suffix, with the Latin or Greek word for the organ of origin as the root. For instance, a cancer of the liver is called *hepatocarcinoma*; a cancer of the fat cells is called *liposarcoma*. For common cancers, the English organ name is used.

For instance, the most common type of breast cancer is called *ductal carcinoma of the breast* or *mammary ductal carcinoma*. Here, the adjective *ductal* refers to the appearance of the cancer under the microscope, resembling normal breast ducts.

Benign tumors (which are not cancers) are named using -oma as a suffix with the organ name as the root. For instance, a benign tumour of the smooth muscle of the uterus is called *leiomyoma* (the common name of this frequent tumour is *fibroid*). Unfortunately, some cancers also use the -oma suffix, examples being melanoma and seminoma.

Signs and Symptoms

Roughly, cancer symptoms can be divided into three groups:

- *Local symptoms*: unusual lumps or swelling (*tumour*), hemorrhage (bleeding), pain and/or ulceration. Compression of surrounding tissues may cause symptoms such as jaundice (yellowing the eyes and skin).
- *Symptoms of metastasis (spreading)*: enlarged lymph nodes, cough and hemoptysis, hepatomegaly (enlarged liver), bone pain, fracture of affected bones and neurological symptoms. Although advanced cancer may cause pain, it is often not the first symptom.
- *Systemic symptoms*: weight loss, poor appetite, fatigue and cachexia (wasting), excessive sweating (night sweats), anemia and specific paraneoplastic phenomena, i.e. specific conditions that are due to an active cancer, such as thrombosis or hormonal changes.

Every symptom in the above list can be caused by a variety of conditions (a list of which is referred to as the differential diagnosis). Cancer may be a common or uncommon cause of each item.

Causes

Cancer is a diverse class of diseases which differ widely in their causes and biology. Any organism, even plants, can acquire cancer. Nearly all known cancers arise gradually, as errors build up in the cancer cell and its progeny. Anything which replicates (living cells) will probabilistically suffer from errors (mutations). Unless error correction and prevention is properly carried out, the errors will survive, and might be passed along to daughter cells. Normally, the body safeguards against cancer via numerous methods, such as: apoptosis, helper molecules (some DNA polymerases), possibly senescence, etc.

However these error-correction methods often fail in small ways, especially in environments that make errors more likely to arise and propagate. For example, such environments can include the presence of disruptive substances called carcinogens, or periodic injury (physical, heat, etc.), or environments that cells did not evolve to withstand, such as hypoxia. Cancer is thus a *progressive* disease, and these progressive errors slowly accumulate until a cell begins to act contrary to its function in the organism.

The errors which cause cancer are often *self-amplifying*, eventually compounding at an exponential rate. For example:

- A mutation in the error-correcting machinery of a cell might cause that cell and its children to accumulate errors more rapidly
- A mutation in signalling (endocrine) machinery of the cell can send error-causing signals to nearby cells
- A mutation might cause cells to become neoplastic, causing them to migrate and disrupt more healthy cells
- Amutation may cause the cell to become immortal, causing them to disrupt healthy cells forever.

Thus cancer often explodes in something akin to a chain reaction caused by a few errors, which compound into more severe errors. Errors which produce more errors are effectively the root cause of cancer, and also the reason that cancer is so hard to treat: even if there were 10,000,000,000 cancerous cells and one killed all but 10 of those cells, those cells (and other error-prone precancerous cells) could still self-replicate or send error-causing signals to other cells, starting the process over again. This rebellion-like scenario is an undesirable survival of the fittest, where the driving forces of evolution work against the body's design and enforcement of order.

In fact, once cancer has begun to develop, this same force continues to drive the progression of cancer towards more invasive stages, and is called clonal evolution.

Research about cancer causes often falls into the following categories:

- Agents (e.g. viruses) and events (e.g. mutations) which cause or facilitate genetic changes in cells destined to become cancer.
- The precise nature of the genetic damage, and the genes which are affected by it.
- The consequences of those genetic changes on the biology of the cell, both in generating the defining properties of a cancer cell, and in facilitating additional genetic events which lead to further progression of the cancer.

Chemicals

Cancer pathogenesis is traceable back to DNA mutations that impact cell growth and metastasis. Substances that cause DNA mutations are known as mutagens, and mutagens that cause cancers are known as carcinogens. Particular substances have been linked to specific types of cancer. Tobacco smoking is associated with many forms of cancer, and causes 90% of lung cancer. Prolonged exposure to asbestos fibres is associated with mesothelioma.

Many mutagens are also carcinogens, but some carcinogens are not mutagens. Alcohol is an example of a chemical carcinogen that is not a mutagen. Such chemicals may promote cancers through stimulating the rate of cell division. Faster rates of replication leaves less time for repair enzymes to repair damaged DNA during DNA replication, increasing the likelihood of a mutation.

Decades of research has demonstrated the link between tobacco use and cancer in the lung, larynx, head, neck, stomach, bladder, kidney, oesophagus and pancreas. Tobacco smoke contains over fifty known carcinogens, including nitrosamines and polycyclic aromatic hydrocarbons.

Tobacco is responsible for about one in three of all cancer deaths in the developed world, and about one in five worldwide. Indeed, lung cancer death rates in the United States have mirrored smoking patterns, with increases in smoking followed by dramatic increases in lung cancer death rates and, more recently, decreases in smoking followed by decreases in lung cancer death rates in men.

However, the numbers of smokers worldwide is still rising, leading to what some organizations have described as the *tobacco epidemic*.

Ionizing Radiation

Sources of ionizing radiation, such as radon gas, can cause cancer. Prolonged exposure to ultraviolet radiation from the sun can lead to melanoma and other skin malignancies. It is estimated that 2% of future cancers will be due to current CT scans.

Non-ionizing radio frequency radiation from mobile phones and other similar RF sources has also been proposed as a cause of cancer, but there is currently little established evidence of such a link.

Infection

Some cancers can be caused by infection. This is especially true in animals such as birds, but also in humans, with viruses responsible for up to 20% of human cancers worldwide. These include human papillomavirus (cervical carcinoma), human polyomaviruses (mesothelioma, brain tumors), Epstein-Barr virus (B-cell lymphoproliferative disease and nasopharyngeal carcinoma), Kaposi's sarcoma herpesvirus (Kaposi's Sarcoma and primary effusion lymphomas), hepatitis B and hepatitis C viruses (hepatocellular carcinoma), Human T-cell leukemia virus-1 (T-cell leukemias), and Helicobacter pylori (gastric carcinoma).

Experimental and epidemiological data imply a causative role for viruses and they appear to be the second most important risk factor for cancer development in humans, exceeded only by tobacco usage. The mode of virally induced tumors can be divided into two, *acutely transforming* or *slowly transforming*. In acutely transforming viruses, the virus carries an overactive oncogene called viral-oncogene (v-onc), and the infected cell is transformed as soon as v-onc is expressed.

In contrast, in slowly transforming viruses, the virus genome is inserts near a proto-oncogene in the host genome. The viral promoter or other transcription regulation elements then cause overexpression of that proto-oncogene. This induces uncontrolled cell division. Because the site of insertion is not specific to proto-oncogenes and the chance of insertion near any proto-oncogene is low, slowly transforming viruses will cause tumors much longer after infection than the acutely transforming viruses.

Hepatitis viruses, including hepatitis B and hepatitis C, can induce a chronic viral infection that leads to liver cancer in 0.47% of hepatitis B patients per year (especially in Asia, less so in North America), and in 1.4% of hepatitis C carriers per year. Liver cirrhosis, whether from chronic viral hepatitis infection or alcoholism, is associated with the development of liver cancer, and the combination of cirrhosis and viral hepatitis presents the highest risk of liver cancer development. Worldwide, liver cancer is one of the most common, and most deadly, cancers due to a huge burden of viral hepatitis transmission and disease.

Advances in cancer research have made a vaccine designed to prevent cancer available. In 2006, the U.S. Food and Drug Administration approved a human papilloma virus vaccine, called Gardasil. The vaccine protects against four HPV types, which together cause 70% of cervical cancers and 90% of genital warts. In March 2007, the US Centres for Disease Control and Prevention (CDC) Advisory Committee on Immunization Practices (ACIP) officially recommended that females aged 11–12 receive the vaccine, and indicated that females as young as age 9 and as old as age 26 are also candidates for immunization.

In addition to viruses, researchers have noted a connection between bacteria and certain cancers. The most prominent example is the link between chronic infection of the wall of the stomach with *Helicobacter pylori* and gastric cancer. Although only a minority of those infected with *Helicobacter* go on to develop cancer, since this pathogen is quite common it is probably responsible for most of these cancers.

Hormonal Imbalances

Some hormones can act in a similar manner to non-mutagenic carcinogens in that they may stimulate excessive cell growth.

Immune System Dysfunction

HIV is associated with a number of malignancies, including Kaposi's sarcoma, non-Hodgkin's lymphoma, and HPV-associated malignancies such as anal cancer and cervical cancer. AIDS-defining illnesses have long included these diagnoses. The increased incidence of malignancies in HIV patients points to the breakdown of immune surveillance as a possible etiology of cancer. Certain other immune deficiency states (e.g. common variable immunodeficiency and IgA deficiency) are also associated with increased risk of malignancy.

Heredity

Most forms of cancer are *sporadic*, meaning that there is no inherited cause of the cancer. There are, however, a number of recognised syndromes where there is an inherited predisposition to cancer, often due to a defect in a gene that protects against tumour formation. Famous examples are:

- certain inherited mutations in the genes *BRCA1* and *BRCA2* are associated with an elevated risk of breast cancer and ovarian cancer

- tumors of various endocrine organs in multiple endocrine neoplasia (MEN types 1, 2a, 2b)
- Li-Fraumeni syndrome (various tumors such as osteosarcoma, breast cancer, soft tissue sarcoma, brain tumors) due to mutations of p53
- Turcot syndrome (brain tumors and colonic polyposis)
- Familial adenomatous polyposis an inherited mutation of the *APC* gene that leads to early onset of colon carcinoma.
- Hereditary nonpolyposis colorectal cancer (HNPCC, also known as Lynch syndrome) can include familial cases of colon cancer, uterine cancer, gastric cancer, and ovarian cancer, without a preponderance of colon polyps.
- Retinoblastoma, when occurring in young children, is due to a hereditary mutation in the retinoblastoma gene.
- Down syndrome patients, who have an extra chromosome 21, are known to develop malignancies such as leukemia and testicular cancer, though the reasons for this difference are not well understood.

Other Causes

Excepting the rare transmissions that occur with pregnancies and only a marginal few organ donors, cancer is generally not a transmissible disease. The main reason for this is tissue graft rejection caused by MHC incompatibility. In humans and other vertebrates, the immune system uses MHC antigens to differentiate between "self" and "non-self" cells because these antigens are different from person to person. When non-self antigens are encountered, the immune system reacts against the appropriate cell. Such reactions may protect against tumour cell engraftment by eliminating implanted cells.

In the United States, approximately 3,500 pregnant women have a malignancy annually, and transplacental transmission of acute leukaemia, lymphoma, melanoma and carcinoma from mother to fetus has been observed. The development of donor-derived tumors from organ transplants is exceedingly rare. The main cause of organ transplant associated tumors seems to be malignant melanoma, that was undetected at the time of organ harvest. though other cases exist.

In fact, cancer from one organism will usually grow in another organism of that species, as long as they share the same histocompatibility genes, proven using mice; however this would never happen in a real-world setting except as described above. In non-humans, a few types of transmissible cancer have been described, wherein the cancer spreads between animals by transmission of the tumour cells themselves.

This phenomenon is seen in dogs with Sticker's sarcoma, also known as canine transmissible venereal tumour, as well as Devil facial tumour disease in Tasmanian devils.

Pathophysiology

Cancer is fundamentally a disease of regulation of tissue growth. In order for a normal cell to transform into a cancer cell, genes which regulate cell growth and differentiation must be altered. Genetic changes can occur at many levels, from gain or loss of entire chromosomes to a mutation affecting a single DNA nucleotide.

There are two broad categories of genes which are affected by these changes. Oncogenes may be normal genes which are expressed at inappropriately high levels, or altered genes which have novel properties.

In either case, expression of these genes promotes the malignant phenotype of cancer cells. Tumour suppressor genes are genes which inhibit cell division, survival, or other properties of cancer cells. Tumour suppressor genes are often disabled by cancer-promoting genetic changes. Typically, changes in many genes are required to transform a normal cell into a cancer cell.

There is a diverse classification scheme for the various genomic changes which may contribute to the generation of cancer cells. Most of these changes are mutations, or changes in the nucleotide sequence of genomic DNA. Aneuploidy, the presence of an abnormal number of chromosomes, is one genomic change which is not a mutation, and may involve either gain or loss of one or more chromosomes through errors in mitosis.

Large-scale mutations involve the deletion or gain of a portion of a chromosome. Genomic amplification occurs when a cell gains many copies (often 20 or more) of a small chromosomal locus, usually containing one or more oncogenes and adjacent genetic material. Translocation occurs when two separate chromosomal regions become abnormally fused, often at a characteristic location. A well-known example of this is the Philadelphia chromosome, or translocation of chromosomes 9 and 22, which occurs in chronic myelogenous leukemia, and results in production of the BCR-abl fusion protein, an oncogenic tyrosine kinase.

Small-scale mutations include point mutations, deletions, and insertions, which may occur in the promoter of a gene and affect its expression, or may occur in the gene's coding sequence and alter the function or stability of its protein product. Disruption of a single gene may also result from integration of genomic material from a DNA virus or retrovirus, and such an event may also result in the expression of viral oncogenes in the affected cell and its descendants.

Prevention

Cancer prevention is defined as active measures to decrease the incidence of cancer. Greater than 30% of cancer is preventable via avoiding risk factors including: tobacco, overweight or obesity, low fruit and vegetable intake, physical inactivity, alcohol, sexually transmitted infection, air pollution.

This can be accomplished by avoiding carcinogens or altering their metabolism, pursuing a lifestyle or diet that modifies cancer-causing factors and/or medical intervention (chemoprevention, treatment of pre-malignant lesions).

The epidemiological concept of "prevention" is usually defined as either primary prevention, for people who have not been diagnosed with a particular disease, or secondary prevention, aimed at reducing recurrence or complications of a previously diagnosed illness.

But the EPIC study published in 2010, tracking the eating habits of 478,000 Europeans suggested that consuming lots of fruits and vegetables has little if any effect on preventing cancer.

Modifiable Factors

The vast majority of cancer risk factors are environmental or lifestyle-related, leading to the claim that cancer is a largely preventable disease. Examples of modifiable cancer risk factors include alcohol consumption (associated with increased risk of oral, esophageal, breast, and other cancers), smoking (80% of women with lung cancer have smoked in the past, and 90% of men), physical inactivity (associated with increased risk of colon, breast, and possibly other cancers), and being overweight/obese (associated with colon, breast, endometrial, and possibly other cancers).

Based on epidemiologic evidence, it is now thought that avoiding excessive alcohol consumption may contribute to reductions in risk of certain cancers; however, compared with tobacco exposure, the magnitude of effect is modest or small and the strength of evidence is often weaker. Other lifestyle and environmental factors known to affect cancer risk (either beneficially or detrimentally) include certain sexually transmitted diseases (such as those conveyed by the human papillomavirus), the use of exogenous hormones, exposure to ionizing radiation and ultraviolet radiation from the sun or from tanning beds, and certain occupational and chemical exposures.

Every year, at least 200,000 people die worldwide from cancer related to their workplace. Millions of workers run the risk of developing cancers such as lung cancer and mesothelioma from inhaling asbestos fibres and tobacco smoke, or leukemia from exposure to benzene at their workplaces. Currently, most cancer deaths caused by occupational risk factors occur in the developed world. It is estimated that approximately 20,000 cancer deaths and 40,000 new cases of cancer each year in the U.S. are attributable to occupation.

Diet

The consensus on diet and cancer is that obesity increases the risk of developing cancer. Particular dietary practices often explain differences in cancer incidence in different countries (e.g. gastric cancer is more common in Japan, while colon cancer is more common in the United States. In this example the preceding consideration of Haplogroups are excluded). Studies have shown that immigrants develop the risk of their new country, often within one generation, suggesting a substantial link between diet and cancer. Whether reducing obesity in a population also reduces cancer incidence is unknown.

Despite frequent reports of particular substances (including foods) having a beneficial or detrimental effect on cancer risk, few of these have an established link to cancer. These reports are often based on studies in cultured cell media or animals. Public health recommendations cannot be made based on these studies until they have been validated in an observational (or occasionally a prospective interventional) trial in humans.

Proposed dietary interventions for primary cancer risk reduction generally gain support from epidemiological association studies. Examples of such studies include reports that reduced meat consumption is associated with decreased risk of colon cancer, and reports that consumption of coffee is associated with a reduced risk of liver cancer.

Studies have linked consumption of grilled meat to an increased risk of stomach cancer, colon cancer, breast cancer, and pancreatic cancer, a phenomenon which could be due to the presence of carcinogens such as benzopyrene in foods cooked at high temperatures. A 2005 secondary prevention study showed that consumption of a plant-based diet and lifestyle changes resulted in a reduction in cancer markers in a group of men with prostate cancer who were using no conventional treatments at the time.

These results were amplified by a 2006 study. Over 2,400 women were studied, half randomly assigned to a normal diet, the other half assigned to a diet containing less than 20% calories from fat. The women on the low fat diet were found to have a markedly lower risk of breast cancer recurrence, in the interim report of December, 2006.

Recent studies have also demonstrated potential links between some forms of cancer and high consumption of refined sugars and other simple carbohydrates. Although the degree of correlation and the degree of causality is still debated, some organizations have in fact begun to recommend reducing intake of refined sugars and starches as part of their cancer prevention regimens.

In November 2007, the American Institute for Cancer Research (AICR), in conjunction with the World Cancer Research Fund (WCRF), published *Food, Nutrition, Physical Activity and the Prevention of Cancer: a Global Perspective*, "the most current and comprehensive analysis of the literature on diet, physical activity and cancer".

The WCRF/AICR Expert Report lists 10 recommendations that people can follow to help reduce their risk of developing cancer, including the following dietary guidelines: (1) reducing intake of foods and drinks that promote weight gain, namely energy-dense foods and sugary drinks, (2) eating mostly foods of plant origin, (3) limiting intake of red meat and avoiding processed meat, (4) limiting consumption of alcoholic beverages, and (5) reducing intake of salt and avoiding mouldy cereals (grains) or pulses (legumes).

Some mushrooms offer an anti-cancer effect, which is thought to be linked to their ability to up-regulate the immune system. Some mushrooms known for this effect include, Reishi, *Agaricus blazei*, Maitake, and *Trametes versicolor*. Research suggests the compounds in medicinal mushrooms most responsible for up-regulating the immune system and providing an anti-cancer effect, are a diverse collection of polysaccharide compounds, particularly beta-glucans. Beta-glucans are known as "biological response modifiers", and their ability to activate the immune system is well documented.

Specifically, beta-glucans stimulate the innate branch of the immune system. Research has shown beta-glucans have the ability to stimulate macrophage, NK cells, T cells, and immune system cytokines. The mechanisms in which beta-glucans stimulate the immune system is only partially understood. One mechanism in which beta-glucans are able to activate the immune system, is by interacting with the Macrophage-1 antigen (CD18) receptor on immune cells.

Vitamins

Vitamin supplementation has not been proven effective in the prevention of cancer. The components of food are also proving to be more numerous and varied than previously understood, so patients are increasingly advised to consume fruits and vegetables for maximal health benefits.

Vitamin D

Low levels of vitamin D is correlated with increased cancer risk. Whether this relationship is causal is yet to be determined.

Beta Carotene

Beta-carotene supplementation has been found to increase slightly, but not significantly risks of lung cancer.

Folic Acid

Folic acid supplementation has not been found effective in preventing colon cancer and may increase colon polyps.

Chemoprevention

The concept that medications could be used to prevent cancer is an attractive one, and many high-quality clinical trials support the use of such chemoprevention in defined circumstances.

Daily use of tamoxifen, a selective estrogen receptor modulator (SERM), typically for 5 years, has been demonstrated to reduce the risk of developing breast cancer in high-risk women by about 50%. A recent study reported that the selective estrogen receptor modulator raloxifene has similar benefits to tamoxifen in preventing breast cancer in high-risk women, with a more favourable side effect profile.

Raloxifene is a SERM like tamoxifen; it has been shown (in the STAR trial) to reduce the risk of breast cancer in high-risk women equally as well as tamoxifen. In this trial, which studied almost 20,000 women, raloxifene had fewer side effects than tamoxifen, though it did permit more DCIS to form.

Finasteride, a 5-alpha-reductase inhibitor, has been shown to lower the risk of prostate cancer, though it seems to mostly prevent low-grade tumors. The effect of COX-2 inhibitors such as rofecoxib and celecoxib upon the risk of colon polyps have been studied in familial adenomatous polyposis patients and in the general population. In both groups, there were significant reductions in colon polyp incidence, but this came at the price of increased cardiovascular toxicity.

Genetic Testing

Genetic testing for high-risk individuals is already available for certain cancer-related genetic mutations. Carriers of genetic mutations that increase risk for cancer incidence can undergo enhanced surveillance, chemoprevention, or risk-reducing surgery. Early identification of inherited genetic risk for cancer, along with cancer-preventing interventions such as surgery or enhanced surveillance, can be lifesaving for high-risk individuals.

Vaccination

Prophylactic vaccines have been developed to prevent infection by oncogenic infectious agents such as viruses, and therapeutic vaccines are in development to stimulate an immune response against cancer-specific epitopes. As reported above, a preventive human papillomavirus vaccine exists that targets certain sexually transmitted strains of human papillomavirus that are associated with the development of cervical cancer and genital warts. The only two HPV vaccines on the market as of October 2007 are Gardasil and Cervarix. There is also a hepatitis B vaccine, which prevents infection with the hepatitis B virus, an infectious agent that can cause liver cancer. A canine melanoma vaccine has also been developed.

Screening

Cancer screening is an attempt to detect unsuspected cancers in an asymptomatic population. Screening tests suitable for large numbers of healthy people must be relatively affordable, safe, noninvasive procedures with acceptably low rates of false positive results. If signs of cancer are detected, more definitive and invasive follow up tests are performed to confirm the diagnosis.

Screening for cancer can lead to earlier diagnosis in specific cases. Early diagnosis may lead to extended life, but may also falsely prolong the lead time to death through lead time bias or length time bias.

A number of different screening tests have been developed for different malignancies. Breast cancer screening can be done by breast self-examination, though this approach was discredited by a 2005 study in over 300,000 Chinese women. Screening for breast cancer with mammograms has been shown to reduce the average stage of diagnosis of breast cancer in a population. Stage of diagnosis in a country has been shown to decrease within ten years of introduction of mammographic screening programs. Colorectal cancer can be detected through fecal occult blood testing and colonoscopy, which reduces both colon cancer incidence and mortality, presumably through the detection and removal of pre-malignant polyps. Similarly, cervical cytology testing (using the Pap smear) leads to the identification and excision of precancerous lesions. Over time, such testing has been followed by a dramatic reduction of cervical cancer incidence and mortality. Testicular self-examination is recommended for men beginning at the age of 15 years to detect testicular cancer. Prostate cancer can be screened using a digital rectal exam along with prostate specific antigen (PSA) blood testing, though some authorities (such as the US Preventive Services Task Force) recommend against routinely screening all men.

Screening for cancer is controversial in cases when it is not yet known if the test actually saves lives. The controversy arises when it is not clear if the benefits of screening outweigh the risks of follow-up diagnostic tests and cancer treatments. For example: when screening for prostate cancer, the PSA test may detect small cancers that would never become life threatening, but once detected will lead to treatment. This situation, called overdiagnosis, puts men at risk for complications from unnecessary treatment such as surgery or radiation. Follow up procedures used to diagnose prostate cancer (prostate biopsy) may cause side effects, including bleeding and infection. Prostate cancer treatment may cause incontinence (inability to control urine flow) and erectile dysfunction (erections inadequate for intercourse). Similarly, for breast cancer, there have recently been criticisms that breast screening programs in some countries cause more problems than they solve. This is because screening of women in the general population will result in a large number of women with false positive results which require extensive follow-up investigations to exclude cancer, leading to having a high number-to-treat (or number-to-screen) to prevent or catch a single case of breast cancer early.

Cervical cancer screening via the Pap smear has the best cost-benefit profile of all the forms of cancer screening from a public health perspective as, largely caused by a virus, it has clear risk factors (sexual contact), and the natural progression of cervical cancer is that it normally spreads slowly over a number of years therefore giving more time for the screening program to catch it early. Moreover, the test is easy to perform and relatively cheap.

For these reasons, it is important that the benefits and risks of diagnostic procedures and treatment be taken into account when considering whether to undertake cancer screening.

Use of medical imaging to search for cancer in people without clear symptoms is similarly marred with problems. There is a significant risk of detection of what has been recently called an *incidentaloma* - a benign lesion that may be interpreted as a malignancy and be subjected to potentially dangerous investigations. Recent studies of CT scan-based screening for lung cancer in smokers have had equivocal results, and systematic screening is not recommended as of July 2007. Randomized clinical trials of plain-film chest X-rays to screen for lung cancer in smokers have shown no benefit for this approach. Canine cancer detection has shown promise, but is still in the early stages of research.

Diagnosis

Most cancers are initially recognized either because signs or symptoms appear or through screening. Neither of these lead to a definitive diagnosis, which usually requires the opinion of a pathologist, a type of physician (medical doctor) who specializes in the diagnosis of cancer and other diseases. People with suspected cancer are investigated with medical tests. These commonly include blood tests, X-rays, CT scans and endoscopy.

Pathology

A cancer may be suspected for a variety of reasons, but the definitive diagnosis of most malignancies must be confirmed by histological examination of the cancerous cells by a pathologist. Tissue can be obtained from a biopsy or surgery. Many biopsies (such as those of the skin, breast or liver) can be done in a doctor's office. Biopsies of other organs are performed under anesthesia and require surgery in an operating room.

The tissue diagnosis given by the pathologist indicates the type of cell that is proliferating, its histological grade, genetic abnormalities, and other features of the tumour. Together, this information is useful to evaluate the prognosis of the patient and to choose the best treatment. Cytogenetics and immunohistochemistry are other types of testing that the pathologist may perform on the tissue specimen.

These tests may provide information about the molecular changes (such as mutations, fusion genes, and numerical chromosome changes) that has happened in the cancer cells, and may thus also indicate the future behavior of the cancer (prognosis) and best treatment.

Management

Many management options for cancer exist including: chemotherapy, radiation therapy, surgery, immunotherapy, monoclonal antibody therapy and other methods. Which are used depends upon the location and grade of the tumour and the stage of the disease, as well as the general state of a person's health. Experimental cancer treatments are also under development.

Complete removal of the cancer without damage to the rest of the body is the goal of treatment. Sometimes this can be accomplished by surgery, but the propensity of cancers to invade adjacent tissue or to spread to distant sites by microscopic metastasis often limits its effectiveness. Surgery often required the removal of a wide surgical margin or a free margin.

The width of the free margin depends on the type of the cancer, the method of removal (CCPDMA, Mohs surgery, POMA, etc.). The margin can be as little as 1 mm for basal cell cancer using CCPDMA or Mohs surgery, to several centimeters for aggressive cancers. The effectiveness of chemotherapy is often limited by toxicity to other tissues in the body. Radiation can also cause damage to normal tissue.

Because "cancer" refers to a class of diseases, it is unlikely that there will ever be a single "cure for cancer" any more than there will be a single treatment for all infectious diseases. Angiogenesis inhibitors were once thought to have potential as a "silver bullet" treatment applicable to many types of cancer, but this has not been the case in practice.

Prognosis

Cancer has a reputation as a deadly disease. While this certainly applies to certain particular types, the truths behind the historical connotations of cancer are increasingly overturned by advances in medical care. Some types of cancer have a prognosis that is substantially better than nonmalignant diseases such as heart failure and stroke.

Progressive and disseminated malignant disease has a substantial impact on a cancer patient's quality of life, and many cancer treatments (such as chemotherapy) may have severe side-effects. In the advanced stages of cancer, many patients need extensive care, affecting family members and friends. Palliative care solutions may include permanent or "respite" hospice nursing.

Emotional Impact

Many local organizations offer a variety of practical and support services to people with cancer. Support can take the form of support groups, counseling, advice, financial assistance, transportation to and from treatment, films or information about cancer. Neighborhood organizations, local health care providers, or area hospitals may have resources or services available. Counseling can provide emotional support to cancer patients and help them better understand their illness. Different types of counseling include individual, group, family, peer counseling, bereavement, patient-to-patient, and sexuality.

Many governmental and charitable organizations have been established to help patients cope with cancer. These organizations are often involved in cancer prevention, cancer treatment, and cancer research.

Epidemiology

As of 2004, worldwide cancer caused 13% of all deaths (7.4 million). The leading causes were: lung cancer (1.3 million deaths/year), stomach cancer (803,000 deaths), colorectal cancer (639,000 deaths), liver cancer (610,000 deaths), and breast cancer (519,000 deaths). Greater than 30% of cancer is preventable via avoiding risk factors including: tobacco, overweight or obesity, low fruit and vegetable intake, physical inactivity, alcohol, sexually transmitted infections, and air pollution.

In the United States, cancer is responsible for 25% of all deaths with 30% of these from lung cancer. The most commonly occurring cancer in men is prostate cancer (about 25% of new cases) and in women is breast cancer (also about 25%). Cancer can occur in children and adolescents, but it is uncommon (about 150 cases per million in the U.S.), with leukemia the most common. In the first year of life the incidence is about 230 cases per million in the U.S., with the most common being neuroblastoma.

In the developed world, one in three people will develop cancer during their lifetimes. If *all* cancer patients survived and cancer occurred *randomly*, the lifetime odds of developing an second primary cancer would be one in nine. However, cancer survivors have an increased risk of developing a second primary cancer, and the odds are about two in nine. About half of these second primaries can be attributed to the normal one-in-nine risk associated with random chance. The increased risk is believed to be primarily due to the same risk factors that produced the first cancer, such as the person's genetic profile, alcohol and tobacco use, obesity, and environmental exposures, and partly due, to the treatment for the first cancer, which typically includes mutagenic chemotherapeutic drugs or radiation. Cancer survivors may also be more likely to comply with recommended screening, and thus may be more likely than average to detect cancers.

History

Today, the Greek term carcinoma is the medical term for a malignant tumour derived from epithelial cells. It is Celsus who translated *carcinos* into the Latin *cancer*, also meaning crab. Galen used "*oncos*" to describe *all* tumours, the root for the modern word oncology.

Hippocrates described several kinds of cancers. He called benign tumours *oncos*, Greek for swelling, and malignant tumours *carcinos*, Greek for crab or crayfish. This name comes from the appearance of the cut surface of a solid malignant tumour, with "the veins stretched on all sides as the animal the crab has its feet, whence it derives its name". He later added the suffix *-oma*, Greek for swelling, giving the name *carcinoma*. Since it was against Greek tradition to open the body, Hippocrates only described and made drawings of outwardly visible tumors on the skin, nose, and breasts. Treatment was based on the humor theory of four bodily fluids (black and yellow bile, blood, and phlegm). According to the patient's humor, treatment consisted of diet, blood-letting, and/or laxatives. Through the centuries it was discovered that cancer could occur anywhere in the body, but humor-theory based treatment remained popular until the 19th century with the discovery of cells.

The oldest known description and surgical treatment of cancer was discovered in Egypt and dates back to approximately 1600 B.C. The Papyrus describes 8 cases of ulcers of the breast that were treated by cauterization, with a tool called "the fire drill." The writing says about the disease, "There is no treatment."

Another very early surgical treatment for cancer was described in the 1020s by Avicenna (Ibn Sina) in *The Canon of Medicine*. He stated that the excision should be radical and that all diseased tissue should be removed, which included the use of amputation or the removal of veins running in the direction of the tumour. He also recommended the use of cauterization for the area treated if necessary.

In the 16th and 17th centuries, it became more acceptable for doctors to dissect bodies to discover the cause of death. The German professor Wilhelm Fabry believed that breast cancer was caused by a milk clot in a mammary duct. The Dutch professor Francois de la Boe Sylvius, a follower of Descartes, believed that all disease was the outcome of chemical processes, and that acidic lymph fluid was the cause of cancer. His contemporary Nicolaes Tulp believed that cancer was a poison that slowly spreads, and concluded that it was contagious.

The first cause of cancer was identified by British surgeon Percivall Pott, who discovered in 1775 that cancer of the scrotum was a common disease among chimney sweeps. The work of other individual physicians led to various insights, but when physicians started working together they could make firmer conclusions.

With the widespread use of the microscope in the 18th century, it was discovered that the 'cancer poison' spread from the primary tumour through the lymph nodes to other sites ("metastasis"). This view of the disease was first formulated by the English surgeon Campbell De Morgan between 1871 and 1874. The use of surgery to treat cancer had poor results due to problems with hygiene.

The renowned Scottish surgeon Alexander Monro saw only 2 breast tumour patients out of 60 surviving surgery for two years. In the 19th century, asepsis improved surgical hygiene and as the survival statistics went up, surgical removal of the tumour became the primary treatment for cancer. With the exception of William Coley who in the late 1800s felt that the rate of cure after surgery had been higher *before* asepsis (and who injected bacteria into tumors with mixed results), cancer treatment became dependent on the individual art of the surgeon at removing a tumour. During the same period, the idea that the body was made up of various tissues, that in turn were made up of millions of cells, laid rest the humor-theories about chemical imbalances in the body. The age of cellular pathology was born.

The genetic basis of cancer was recognised in 1902 by the German zoologist Theodor Boveri, professor of zoology at Munich and later in Würzburg. He discovered a method to generate cells with multiple copies of the centrosome, a structure he discovered and named. He postulated that chromosomes were distinct and transmitted different inheritance factors. He suggested that mutations of the chromosomes could generate a cell with unlimited growth potential which could be passed onto its descendants. He proposed the existence of cell cycle check points, tumour suppressor genes and oncogenes. He speculated that cancers might be caused or promoted by radiation, physical or chemical insults or by pathogenic microorganisms.

When Marie Curie and Pierre Curie discovered radiation at the end of the 19th century, they stumbled upon the first effective non-surgical cancer treatment. With radiation also came the first signs of multi-disciplinary approaches to cancer treatment. The surgeon was no longer operating in isolation, but worked together with hospital radiologists to help patients. The complications in communication this brought, along with the necessity of the patient's treatment in a hospital facility rather than at home, also created a parallel process of compiling patient data into hospital files, which in turn led to the first statistical patient studies.

A founding paper of cancer epidemiology was the work of Janet Lane-Claypon, who published a comparative study in 1926 of 500 breast cancer cases and 500 control patients of the same background and lifestyle for the British Ministry of Health. Her ground-breaking work on cancer epidemiology was carried on by Richard Doll and Austin Bradford Hill, who published "Lung Cancer and Other Causes of Death In Relation to Smoking. A Second Report on the Mortality of British Doctors" followed in 1956 (otherwise known as the British doctors study). Richard Doll left the London Medical Research Centre (MRC), to start the Oxford unit for Cancer epidemiology in 1968. With the use of computers, the unit was the first to compile large amounts of cancer data. Modern epidemiological methods are closely linked to current concepts of disease and public health policy. Over the past 50 years, great efforts have been spent on gathering data across medical practise, hospital, provincial, state, and even country boundaries to study the interdependence of environmental and cultural factors on cancer incidence.

Cancer patient treatment and studies were restricted to individual physicians' practices until World War II, when medical research centres discovered that there were large international differences in disease incidence. This insight drove national public health bodies to make it possible to compile health data across practises and hospitals, a process that many countries do today. The Japanese medical community observed that the bone marrow of victims of the atomic bombings of Hiroshima and Nagasaki was completely destroyed. They concluded that diseased bone marrow could also be destroyed with radiation, and this led to the discovery of bone marrow transplants for leukemia. Since World War II, trends in cancer treatment are to improve on a micro-level the existing treatment methods, standardize them, and globalize them to find cures through epidemiology and international partnerships.

Research

Cancer research is the intense scientific effort to understand disease processes and discover possible therapies. The improved understanding of molecular biology and cellular biology due to cancer research has led to a number of new, effective treatments for cancer since President Nixon declared "War on Cancer" in 1971.

Since 1971 the United States has invested over $200 billion on cancer research; that total includes money invested by public and private sectors and foundations. Despite this substantial investment, the country has seen a five percent decrease in the cancer death rate (adjusting for size and age of the population) between 1950 and 2005.

Leading cancer research organizations and projects include the American Association for Cancer Research, the American Cancer Society (ACS), the American Society of Clinical Oncology, the European Organisation for Research and Treatment of Cancer, the National Cancer Institute, the National Comprehensive Cancer Network, and The Cancer Genome Atlas project at the NCI.

2

Genetic Engineering or Recombinant DNA Technology

Genetic engineering, also called genetic modification, is the direct human manipulation of an organism's genetic material in a way that does not occur under natural conditions. It involves the use of recombinant DNA techniques, but does not include traditional animal and plant breeding or mutagenesis. Any organism that is generated using these techniques is considered to be a genetically modified organism.

The first organisms genetically engineered were bacteria in 1973 and then mice in 1974. Insulin producing bacteria were commercialized in 1982 and genetically modified food has been sold since 1994. The most common form of genetic engineering involves the insertion of new genetic material at an unspecified location in the host genome. This is accomplished by isolating and copying the genetic material of interest, generating a construct containing all the genetic elements for correct expression, and then inserting this construct into the host organism. Other forms of genetic engineering include gene targeting and knocking out specific genes via engineered nucleases such as zinc finger nucleases or engineered homing endonucleases.

Genetic engineering techniques have been applied in numerous fields including research, biotechnology, and medicine. Medicines such as insulin and human growth hormone are now produced in bacteria, experimental mice such as the oncomouse and the knockout mouse are being used for research purposes and insect resistant and/or herbicide tolerant crops have been commercialized. Genetically engineered plants and animals capable of producing biotechnology drugs more cheaply than current methods (called pharming) are also being developed and in 2009 the FDA approved the sale of the pharmaceutical protein antithrombin produced in the milk of genetically engineered goats.

Definition

Genetic engineering alters the genetic makeup of an organism using techniques that introduce heritable material prepared outside the organism either directly into the host or into a cell that is then fused or hybridized with the host. This involves using recombinant nucleic acid (DNA or RNA) techniques to form new combinations of heritable genetic material followed by the incorporation of that material either indirectly through a vector system or directly through micro-injection, macro-injection and micro-encapsulation techniques.

Genetic engineering does not include traditional animal and plant breeding, in vitro fertilisation, induction of polyploidy, mutagenesis and cell fusion techniques that do not use recombinant nucleic acids or a genetically modified organism in the process. Cloning and stem cell research, although not considered genetic engineering, are closely related and genetic engineering can be used within them. Synthetic biology is an emerging discipline that takes genetic engineering a step further by introducing artificially synthesized genetic material from raw materials into an organism.

If genetic material from another species is added to the host, the resulting organism is called transgenic. If genetic material from the same species or a species that can naturally breed with the host is used the resulting organism is called cisgenic. Genetic engineering can also be used to remove genetic material from the target organism, creating a knock out organism. In Europe genetic modification is synonymous with genetic engineering while within the United States of America it can also refer to conventional breeding methods.

History

Humans have altered the genomes of species for thousands of years through artificial selection and more recently mutagenesis. Genetic engineering as the direct manipulation of DNA by humans outside breeding and mutations has only existed since the 1970s. The term "genetic engineering" was first coined by Jack Williamson in his science fiction novel *Dragon's Island*, published in 1951, one year before DNA's role in heredity was confirmed by Alfred Hershey and Martha Chase, and two years before James Watson and Francis Crick showed that the DNA molecule has a double-helix structure.

In 1972 Paul Berg created the first recombinant DNA molecules by combined DNA from the monkey virus SV40 with that of the lambda virus. In 1973 Herbert Boyer and Stanley Cohen created the first transgenic organism by inserting antibiotic resistance genes into the plasmid of an *E. coli* bacterium. A year later Rudolf Jaenisch created a transgenic mouse by introducing foreign DNA into its embryo, making it the world's first transgenic animal.

In 1976 Genentech, the first genetic engineering company was founded by Herbert Boyer and Robert Swanson and a year later and the company produced a human protein (somatostatin) in *E.coli*. Genentech announced the production of genetically engineered human insulin in 1978. In 1980, the U.S. Supreme Court in the Diamond v. Chakrabarty case ruled that genetically altered life could be patented.

The insulin produced by bacteria, branded humulin, was approved for release by the Food and Drug Administration in 1982. The first field trials of genetically engineered plants occurred in France and the USA in 1986, tobacco plants were engineered to be resistant to herbicides. The People's Republic of China was the first country to commercialize transgenic plants, introducing a virus-resistant tobacco in 1992. In 1994 Calgene attained approval to commercially release the Flavr Savr tomato, a tomato engineered to have a longer shelf life. In 1994, the European Union approved tobacco engineered to be resistant to the herbicide bromoxynil, making it the first genetically engineered crop commercialized in Europe.

In 1995, Bt Potato was approved safe by the Environmental Protection Agency, making it the first pesticide producing crop to be approved in the USA. In 2009 11 transgenic crops were grown commercially in 25 countries, the largest of which by area grown were the USA, Brazil, Argentina, India, Canada, China, Paraguay and South Africa.

In 2010, scientists at the J. Craig Venter Institute, announced that they had created the first synthetic bacterial genome, and added it to a cell containing no DNA. The resulting bacterium, named Synthia, was the world's first synthetic life form.

Process

Isolating the Gene

First, the gene to be inserted into the genetically modified organism must be chosen and isolated. Presently, most genes transferred into plants provide protection against insects or tolerance to herbicides. In animals the majority of genes used are growth hormone genes. Once chosen the genes must be isolated. This typically involves multiplying the gene using polymerase chain reaction (PCR). If the chosen gene or the donor organism's genome has been well studied it may be present in a genetic library. If the DNA sequence is known, but no copies of the gene are available, it can be artificially synthesized. Once isolated, the gene is inserted into a bacterial plasmid.

Constructs

The gene to be inserted into the genetically modified organism must be combined with other genetic elements in order for it to work properly. The gene can also be modified at this stage for better expression or effectiveness. As well as the gene to be inserted most constructs contain a promoter and terminator region as well as a selectable marker gene. The promoter region initiates transcription of the gene and can be used to control the location and level of gene expression, while the terminator region ends transcription. The selectable marker, which in most cases confers antibiotic resistance to the organism it is expressed in, is needed to determine which cells are transformed with the new gene. The constructs are made using recombinant DNA techniques, such as restriction digests, ligations and molecular cloning.

Gene Targeting

The most common form of genetic engineering involves inserting new genetic material randomly within the host genome. Other techniques allow new genetic material to be inserted at a specific location in the host genome or generate mutations at desired genomic loci capable of knocking out endogenous genes.

The technique of gene targeting uses homologous recombination to target desired changes to a specific endogenous gene. This tends to occur at a relatively low frequency in plants and animals and generally requires the use of selectable markers. The frequency of gene targeting can be greatly enhanced with the use of engineered nucleases such as zinc finger nucleases, engineered homing endonucleases, or nucleases created from TAL effectors. In addition to enhancing gene targeting, engineered nucleases can also be used to introduce mutations at endogenous genes that generate a gene knockout.

Transformation

About 1% of bacteria are naturally able to take up foreign DNA but it can also be induced in other bacteria. Stressing the bacteria for example, with a heat shock or an electric shock, can make the cell membrane permeable to DNA that may then incorporate into their genome or exist as extrachromosomal DNA. DNA is generally inserted into animal cells using microinjection, where it can be injected through the cells nuclear envelope directly into the nucleus or through the use of viral vectors. In plants the DNA is generally inserted using *Agrobacterium*-mediated recombination or biolistics. In *Agrobacterium*-mediated recombination the plasmid construct must also contain T-DNA. *Agrobacterium* naturally inserts DNA from a tumour inducing plasmid into any susceptible plant's genome it infects, causing crown gall disease. The T-DNA region of this plasmid is responsible for insertion of the DNA. The genes to be inserted are cloned into a binary vector, which contains T-DNA and can be grown in both *E. Coli* and *Agrobacterium*. Once the binary vector is constructed the plasmid is transformed into *Agrobacterium* containing no plasmids and plant cells are infected. The *Agrobacterium* will then naturally insert the genetic material into the plant cells.

In biolistics particles of gold or tungsten are coated with DNA and then shot into young plant cells or plant embryos. Some genetic material will enter the cells and transform them. This method can be used on plants that are not susceptible to *Agrobacterium* infection and also allows transformation of plant plastids. Another transformation method for plant and animal cells is electroporation. Electroporation involves subjecting the plant or animal cell to an electric shock, which can make the cell membrane permeable to plasmid DNA. In some cases the electroporated cells will incorporate the DNA into their genome. Due to the damage caused to the cells and DNA the transformation efficiency of biolistics and electroporation is lower than agrobacterial mediated transformation and microinjection.

Selection

Not all the organism's cells will be transformed with the new genetic material; in most cases a selectable marker is used to differentiate transformed from untransformed cells. If a cell has been successfully transformed with the DNA it will also contain the marker gene. By growing the cells in the presence of an antibiotic or chemical that selects or marks the cells expressing that gene it is possible to separate the transgenic events from the non-transgenic. Another method of screening involves using a DNA probe that will only stick to the inserted gene. A number of strategies have been developed that can remove the selectable marker from the mature transgenic plant.

Regeneration

As often only a single cell is transformed with genetic material the organism must be regrown from that single cell. As bacteria consist of a single cell and reproduce clonally regeneration is not necessary. In plants this is accomplished through the use of tissue culture. Each plant species has different requirements for successful regeneration through tissue culture. If successful an adult plant is produced that contains the transgene in every cell. In animals it is necessary to ensure that the inserted DNA is present in the embryonic stem cells. When the offspring is produced they can be screened for the presence of the gene. All offspring from the first generation will be heterozygous for the inserted gene and must be mated together to produce a homozygous animal.

Confirmation

Further tests using PCR, Southern Blots and Bioassays are needed to confirm that the gene is expressed and functions correctly. The organism's offspring are also tested to ensure that the trait can be inherited and that it follows a Mendelian inheritance pattern.

Applications

Genetic engineering has applications in medicine, research, industry and agriculture and can be used on a wide range of plants, animals and micro organism.

Medicine

In medicine genetic engineering has been used to mass-produce insulin, human growth hormones, follistim (for treating infertility), human albumin, monoclonal antibodies, antihemophilic factors, vaccines and many other drugs. Vaccination generally involves injecting weak live, killed or inactivated forms of viruses or their toxins into the person being immunized. Genetically engineered viruses are being developed that can still confer immunity, but lack the infectious sequences. Mouse hybridomas, cells fused together to create monoclonal antibodies, have been humanised through genetic engineering to create human monoclonal antibodies.

Genetic engineering is used to create animal models of human diseases. Genetically modified mice are the most common genetically engineered animal model. They have been used to study and model cancer (the oncomouse), obesity, heart disease, diabetes, arthritis, substance abuse, anxiety, aging and Parkinson disease. Potential cures can be tested against these mouse models. Also genetically modified pigs have been bred with the aim of increasing the success of pig to human organ transplantation. Gene therapy is the genetic engineering of humans by replacing defective human genes with functional copies. This can occur in somatic tissue or germline tissue. If the gene is inserted into the germline tissue it can be passed down to that person's descendants. Gene therapy has been used to treat patients suffering from immune deficiencies (notably Severe combined immunodeficiency) and trials have been carried out on other genetic disorders. The success of gene therapy so far has been limited and a patient (Jesse Gelsinger) has died during a clinical trial testing a new treatment. There are also ethical concerns should the technology be used not just for treatment, but for enhancement, modification or alteration of a human beings' appearance, adaptability, intelligence, character or behavior. The distinction between cure and enhancement can also be difficult to establish. Transhumanists consider the enhancement of humans desirable.

Research

Genetic engineering is an important tool for natural scientists. Genes and other genetic information from a wide range of organisms are transformed into bacteria for storage and modification, creating genetically modified bacteria in the process. Bacteria are cheap, easy to grow, clonal, multiply quickly, relatively easy to transform and can be stored at -80°C almost indefinitely. Once a gene is isolated it can be stored inside the bacteria providing an unlimited supply for research.

Organisms are genetically engineered to discover the functions of certain genes. This could be the effect on the phenotype of the organism, where the gene is expressed or what other genes it interacts with. These experiments generally involve loss of function, gain of function, tracking and expression.

- Loss of function experiments, such as in a gene knockout experiment, in which an organism is engineered to lack the activity of one or more genes. A knockout experiment involves the creation and manipulation of a DNA construct *in vitro*, which, in a simple knockout, consists of a copy of the desired gene, which has been altered such that it is non-functional. Embryonic stem cells incorporate the altered gene, which replaces the already present functional copy. These stem cells are injected into blastocysts, which are implanted into surrogate mothers. This allows the experimenter to analyze the defects caused by this mutation and thereby determine the role of particular genes. It is used especially frequently in developmental biology. Another method, useful in organisms such as Drosophila (fruit fly), is to induce mutations in a large population and then screen the progeny for the desired mutation. A similar process can be used in both plants and prokaryotes.
- Gain of function experiments, the logical counterpart of knockouts. These are sometimes performed in conjunction with knockout experiments to more finely establish the function of the desired gene. The process is much the same as that in knockout engineering, except that the construct is designed to increase the function of the gene, usually by providing extra copies of the gene or inducing synthesis of the protein more frequently.
- Tracking experiments, which seek to gain information about the localization and interaction of the desired protein. One way to do this is to replace the wild-type gene with a 'fusion' gene, which is a

juxtaposition of the wild-type gene with a reporting element such as green fluorescent protein (GFP) that will allow easy visualization of the products of the genetic modification. While this is a useful technique, the manipulation can destroy the function of the gene, creating secondary effects and possibly calling into question the results of the experiment. More sophisticated techniques are now in development that can track protein products without mitigating their function, such as the addition of small sequences that will serve as binding motifs to monoclonal antibodies.

- Expression studies aim to discover where and when specific proteins are produced. In these experiments, the DNA sequence before the DNA that codes for a protein, known as a gene's promoter, is reintroduced into an organism with the protein coding region replaced by a reporter gene such as GFP or an enzyme that catalyzes the production of a dye. Thus the time and place where a particular protein is produced can be observed. Expression studies can be taken a step further by altering the promoter to find which pieces are crucial for the proper expression of the gene and are actually bound by transcription factor proteins; this process is known as promoter bashing.

Industrial

By engineering genes into bacterial plasmids it is possible to create a biological factory that can produce proteins and enzymes. Some genes do not work well in bacteria, so yeast, a eukaryote, can also be used. Bacteria and yeast factories have been used to produce medicines such as insulin, human growth hormone, and vaccines, supplements such as tryptophan, aid in the production of food (chymosin in cheese making) and fuels. Other applications involving genetically engineered bacteria being investigated involve making the bacteria perform tasks outside their natural cycle, such as cleaning up oil spills, carbon and other toxic waste.

Agriculture

One of the best-known and controversial applications of genetic engineering is the creation of genetically modified food. There are three generations of genetically modified crops. First generation crops have been commercialized and most provide protection from insects and/or resistance to herbicides. There are also fungal and virus resistant crops developed or in development. They have been developed to make the insect and weed management of crops easier and can indirectly increase crop yield. The second generation of genetically modified crops being developed aim to directly improve yield by improving salt, cold or drought tolerance and to increase the nutritional value of the crops.

The third generation consists of pharmaceutical crops, crops that contain edible vaccines and other drugs. Some agriculturally important animals have been genetically modified with growth hormones to increase their size while others have been engineered to express drugs and other proteins in their milk.

The genetic engineering of agricultural crops can increase the growth rates and resistance to different diseases caused by pathogens and parasites. This is beneficial as it can greatly increase the production of food sources with the usage of fewer resources that would be required to host the world's growing populations.

These modified crops would also reduce the usage of chemicals, such as fertilizers and pesticides, and therefore decrease the severity and frequency of the damages produced by these chemical pollution. Ethical and safety concerns have been raised around the use of genetically modified food.

A major safety concern relates to the human health implications of eating genetically modified food, in particular whether toxic or allergic reactions could occur.

Gene flow into related non-transgenic crops, off target effects on beneficial organisms and the impact on biodiversity are important environmental issues. Ethical concerns involve religious issues, corporate control of the food supply, intellectual property rights and the level of labelling needed on genetically modified products.

Other Uses

In materials science, a genetically modified virus has been used to construct a more environmentally friendly lithium-ion battery. Some bacteria have been genetically engineered to create black and white photographs while others have potential to be used as sensors by expressing a fluorescent protein under certain environmental conditions. Genetic engineering is also being used to create BioArt and novelty items such as blue roses, and glowing fish.

Opposition and Criticism

A 2010 study of Canola found transgenes in 80% of wild (uncultivated or "feral") varieties in North Dakota, meaning 80% of the plants which had established themselves in the area were genetically engineered varieties.

The researchers stated that "we found the highest densities of [such transgene-containing] plants near agricultural fields and along major freeways, but we were also finding plants in the middle of nowhere" adding that "over time,..the build-up of different types of herbicide resistance in feral [natural] canola and closely related weeds, like field mustard, could make it more difficult to manage these plants using herbicides."

Human Genetic Engineering

Human genetic engineering is the alteration of an individual's genotype with the aim of choosing the phenotype of a newborn or changing the existing phenotype of a child or adult. It holds the promise of curing genetic diseases like cystic fibrosis, and increasing the immunity of people to viruses.

It is speculated that genetic engineering could be used to change physical appearance, metabolism, and even improve mental faculties like memory and intelligence, although for now these uses seem to be of lower priority to researchers and are therefore limited to science fiction.

History

The first gene therapy trials on humans began in 1990 on patients with Severe Combined Immunodeficiency (SCID). In 2000, the first gene therapy "success" resulted in SCID patients with a functional immune system. These trials were stopped when it was discovered that two of ten patients in one trial had developed leukemia resulting from the insertion of the gene-carrying retrovirus near an oncogene. In 2007, four of the ten patients had developed leukemia. Work is now focusing on correcting the gene without triggering an oncogene. Trial treatments of SCID have been gene therapy's only success; since 1999, gene therapy has restored the immune systems of at least 17 children with two forms (ADA-SCID and X-SCID) of the disorder.

Human genetic engineering is already being used on a small scale to allow infertile women with genetic defects in their mitochondria to have children. Healthy human eggs from a second mother are used. The child produced this way has genetic information from two mothers and one father. The changes made are germline changes and will likely be passed down from generation to generation, and, thus, are a permanent change to the human genome.

Other forms of human genetic engineering are still theoretical. Recombinant DNA research is usually performed to study gene expression and various human diseases. Some drastic demonstrations of gene modification have been made with mice and other animals, however, testing on humans is generally considered off-limits. In some instances changes are usually brought about by removing genetic material from one organism and transferring them into another species.

Methods

Somatic

Somatic genetic engineering involves adding genes to cells other than egg or sperm cells. For example, if a person had a disease caused by a defective gene, a healthy gene could be added to the affected cells to treat the disorder. As of now, this is likely to take the form of gene therapy. The distinguishing characteristic of somatic engineering is that it is non-inheritable, i.e. the new gene would not be passed to the recipient's offspring.

There are two techniques researchers are currently experimenting with:

- Viruses are good at injecting their DNA payload into human cells and reproducing it. By adding the desired DNA to the DNA of non-pathogenic virus, a small amount of virus will reproduce the desired DNA and spread it all over the body.
- Manufacture large quantities of DNA, and somehow package it to induce the target cells to accept it, either as an addition to one of the original 23 chromosomes, or as an independent 24th human artificial chromosome.

Germline

Germline engineering involves changing genes in eggs, sperm, or very early embryos. This type of engineering is inheritable, meaning that the modified genes would appear not only in offspring that resulted from the procedure, but also in subsequent generations.

Uses

Two motivators of human genetic engineering are referred to as "negative" and "positive". The former aims to remove genetic disorders and the latter aims to alter phenotypic expression to result in an enhanced being.

Negative Genetic Engineering (Cures and Treatments)

When treating problems that arise from genetic disorder, one solution is gene therapy, also known as negative genetic engineering.

A genetic disorder is a condition caused by the genetic code of the individual, such as spina bifida or autism. When this happens, genes may be expressed in unfavorable ways or not at all, and this generally leads to further complications.

The idea of gene therapy is that a non-pathogenic virus or other delivery systems can be used to insert into DNA—a good copy of the gene—into cells of the living individual. The modified cells would divide as normal and each division would produce cells that express the desired trait.

The result would be that he/she would then have the ability to express the trait that was previously absent, at least partially. This form of genetic engineering could help alleviate many problems, such as diabetes, cystic fibrosis, or other genetic diseases.

Positive Genetic Engineering (Enhancement)

The potential of genetic engineering to cure medical conditions opens the question of exactly what such a condition is. Some view aging and death as medical conditions and therefore potential targets for engineering solutions. They see human genetic engineering potentially as a key tool in this. The difference between cure and enhancement from this perspective is merely one of degree. Theoretically genetic engineering could be used to drastically change people's genomes, which could enable people to regrow limbs and other organs, perhaps even extremely complex ones such as the spine.

It could also be used to make people smarter, stronger, or to increase the capacity of the lungs, among other things. If a gene exists in nature, perhaps it could be changed into a human cell. In this view, there is no qualitative difference (only a quantitative one) between, for instance, a genetic intervention to cure muscular atrophy, and a genetic intervention to improve muscle function even when those muscles are functioning at or below the human average (since there is also an average muscle function for those with a particular type of dystrophy, which the treatment would improve upon).

Others feel, there is an important distinction between using genetic technologies to treat those who are suffering, and to make those who are already healthy seem more superior to the average person. Though theory and speculation suggest, that genetic engineering could be used to make people stronger, faster, smarter, or to increase lung capacity. The AAAS report finds that there is little evidence to support this theory. Can this currently be done without very unsafe and therefore unethical human experiments. Because different cells have different tasks, changing one cell to do a function differently, will not only affect that one task, but it can affect many other tasks as well.

Controversy

Ethics

The genetic engineering of humans has raised many controversial ethical issues. While negative genetic engineering (gene therapy) does indeed raise a debate, the use of genetic engineering for human enhancement arouses the strongest feelings on both sides.

Genetic engineering is tested on animals, often including primates. Some animal rights activists find this inhumane.

Genetic engineering can be used to cure peoples with diabetes. It is possible to extract genes from cells which are called beta cells and then to insert the insulin producing genes into a bacterium. Then the bacterium will start producing insulin. Genetic modification of embryos can pose an ethical question about the rights of the baby. One belief is that every fetus should be free to not be genetically modified. Others believe that parents hold the rights to change their unborn children. Still others believe that every child should have the right to be born free from preventable diseases.

Molecular Biologist Lee M. Silver believes that unlike Aldous Huxley's Brave New World, where a totalitarian government controls all of the genetic enhancements (they actually use eugenics instead of direct genetic modification) in society, the use of gene therapy to design children will be spread through what he calls "free market eugenics" (Silver 315). Wealthy families will opt to design their child with genetic advantages because other families are doing so, and everybody wants to provide their newborn child with the best opportunities in life, with a leg up on the competition.

The greatest fear for Silver is that we will design so many children with germline gene therapy, that the families wealthy enough to design their children, will pass down these enhanced traits to future generations. This gene therapy will obviously cost money, and the less wealthy families will be left to procreate naturally, and introduce their children into the world disadvantaged from their first breath.

The impact on society will be a new alignment of classes, no longer will we separate people by their ethnic differences, the new division will be between what Silver calls 'the naturals' and 'the GenRich', or genetically enhanced. The major worry here is that the 'genetic gulf' between these two classes will become so wide that humans will become separate species (Silver 313).

In Popular Culture

- *Maximum Ride series by James Patterson:* The main characters are six human children who had bird DNA injected into them while they were in their mothers' wombs.
- *Mobile Suit Gundam SEED (anime):* Set in a world in which genetically modified humans, termed 'Coordinators', have been ostracized and isolated from unmodified humans, termed 'Naturals'. Due to extreme differences in mental and physical abilities between the two groups, racial, economic, and political issues have arisen, culminating in war. Gundam Seed addresses such concerns as animosity caused by the jealousy of Naturals over Coordinator abilities, both groups looking down on one another as being lesser life forms, and genocidal factions emerging on both sides. The series primarily explores these issues from the point of view of a Coordinator protagonist who finds himself fighting on the side of the Naturals, as his childhood friend has become a member of the Coordinator military, giving a perspective on both sides of the conflict.

- *Gattaca (film):* Presents a biopunk vision of a society driven by new eugenics. Children of the middle and upper classes are selected through preimplantation genetic diagnosis to ensure they possess the best hereditary traits of their parents.
- *Oryx and Crake by Margaret Atwood:* Apocalyptic pseudo-dystopian story, one of the chief plot points of which involves the genetic engineering of a new type of transhuman and the pathological decimation of *Homo sapiens*.
- *Bioshock (Video Game):* The main enemies the player encounters over the course of the game are known as splicers, so called for their genetic manipulation, or gene splicing. In the game a compound called ADAM is responsible for the genetic manipulation. The compound is harvested from a species of sea slugs, it acts like a seemingly benign form of cancer, destroying native cells and replacing them with the unstable stem versions. The splicers have all become so addicted to the ADAM that they will kill anything or anyone for it.
- *Old Man's War series by John Scalzi:* To create an army of soldiers capable of defending the human race from endless alien hordes the Colonial Defense Forces recruits 75 year olds and gives them new younger bodies capable of super human feats.
- *Batman Beyond:* In the episode *Splicers* and making appearances later in the series, a new fad of splicing animal genes for cosmetic and enhancement purposes makes the teen scene. When research shows that splicing increases aggression in users, resulting it being banned in Gotham, only finding a place in the criminal underworld later. Notable Splicers are Woof of the Jokerz (spliced with hyena DNA), and the Zander (spliced with T-Rex DNA), leader of KOBRA.
- *Halo (series) (Video Game):* the protagonist and several other characters in the Halo universe known as Spartans have all gone through comprehensive genetic augmentation, making them almost super human. The augmentations vary from Muscular Enhancement Injections to a Catalytic Thyroid Implant. The effects include nearly unbreakable bones, increased reflexes and increased muscle tissue with increased density. The SPARTAN-III program also included a mutagen that increased aggression and injury tolerance.
- *Deus Ex:* The protagonists of both games in the series are genetically modified by injections called nano-augs (Deus Ex) and bio mods (Invisible War). These nanite injections alter the host's genes to enhance them with new skill that would help them through different obstacles in the

game. Both the protagonists and antagonists are bio-modified along with other support characters and neutral characters.

- *PROTOTYPE:* The story is centered around a genetic engineering company called Gentek. The engineers in Gentek modified a chimeran-virus to make it 10 times as deadly, to 'unlock' previously dormant parts of a subjects DNA, and they made it to the point of where it could copy those infected with it down to the genetic level. The new 'infected' being then has complete and total control over its genetic structure, allowing it to develop instantaneous shape-shifting abilities. However, the game reveals that the only 'infected' being that can do this is a person already dead (or dying) when the virus enters the bloodstream. The virus must also be in a state of emergency to survive and not die out.
- *Warhammer 40,000:* In the games, and table-top series, Space marines are the result of Genetic Engineering, by replacing organs, and limbs that have been engineered. The Marine himself, also goes through rigorous physical and mental training, to make him accustom to the newly added appendages and organs.

GM Food Controversy

The genetically modified foods controversy is a dispute over the relative advantages and disadvantages of genetically modified (GM) food crops and other uses of genetically-modified organisms in food production. The dispute involves biotechnology companies, governmental regulators, non-governmental organizations and scientists. The dispute is most intense in Japan and Europe, where public concern about GM food is higher than in other parts of the world such as the United States.

In the United States GM crops are more widely grown and the introduction of these products has been less controversial. The five key areas of political controversy related to genetically engineered food are food safety, the effect on natural ecosystems, gene flow into non GE crops, moral/religious concerns, and corporate control of the food supply. To date, not a single instance of harm to human health or the environment has been documented with GM crops. Several benefits have been widely accepted and are uncontested in the scientific literature.

These include reductions in insecticide use on GE cotton, enhanced biological diversity in GE cotton fields (compared to non-GE fields), enhanced farmer income and communal benefits, increased yields for poor farmers and improved health of farmworkers. Although the use of herbicide tolerant crops remain controversial, because of the need to spray herbicides, it is clear that the use of these crops has promoted a shift to less toxic herbicides.

The concept of substantial equivalence has been established to address the fears of some consumers about the safety of GE crops. "Substantial equivalence embodies the concept that if a new food or food component is found to be substantially equivalent to an existing food or food component, it can be treated in the same manner with respect to safety (i.e., the food or food component can be concluded to be as safe as the conventional food or food component)" (Joint FAO/WHO Biotechnology and Food Safety Report, 1996).

The rationale for this approach is that it would be impossible to test all the new crop varieties every year for food safety. Only a few food crops on the market have been shown to cause adverse health effects and all of these were the result of conventional genetic modification, not from genetic engineering To date no adverse health effects caused by products approved for sale have been documented, although two products failed initial safety testing and were discontinued, due to allergic reactions.

Most feeding trials have observed no toxic effects and saw that GM foods were equivalent in nutrition to unmodified foods, although a few non-peer-reviewed reports speculate physiological changes to GM food. Although there is now broad scientific consensus that GE crops on the market are safe to eat, some scientists and advocacy groups such as Greenpeace and World Wildlife Fund call for additional and more rigorous testing before marketing genetically engineered food.

Another area of controversy is what effect pest and herbicide-resistant crops have on ecosystems. For example, whereas 14 years of studies indicate cotton engineered for resistance to insects has resulted in massive reduction in insecticides globally, the use of crops engineered for herbicide tolerance has changed herbicide patterns, with a clear shift to use of less toxic herbicides. Studies assessing diversity of non-target insects has revealed enhanced biological diversity in GE fields compared to non-GE fields.

The risk and effects of have also been cited as concerns, with the possibility that genes might spread from modified crops to wild relatives. There is no evidence that the risk of pollen transfer is greater for any of the GE crops on the market as compared to non-GE crops.

Health Risks and Benefits

Present Knowledge on GM Food Safety

Worldwide, there is a range of perspectives within non-governmental organizations on the safety of GM foods. For example, the US pro-GM group AgBioWorld has argued that GM foods have been proven safe, while other pressure groups and consumer rights groups, such as the Organic Consumers

Association, and Greenpeace claim the long-term health risks which GM could pose, or the environmental risks associated with GM, have not yet been adequately investigated. In Japan, Consumers Union of Japan are opposed to GMO foods. They also claim that truly independent research in these areas is systematically blocked by the GM corporations which own the GM seeds and reference materials.

A 2008 review published by the Royal Society of Medicine noted that GM foods have been eaten by millions of people worldwide for over 15 years, with no reports of ill effects. Similarly a 2004 report from the US National Academies of Sciences stated: "To date, no adverse health effects attributed to genetic engineering have been documented in the human population." A 2004 review of feeding trials in the *Italian Journal of Animal Science* found no differences among animals eating genetically modified plants.

A 2005 review in *Archives of Animal Nutrition* concluded that first-generation genetically modified foods had been found to be similar in nutrition and safety to non-GM foods, but noted that second-generation foods with "significant changes in constituents" would be more difficult to test, and would require further animal studies.

However, a 2009 review in *Nutrition Reviews* found that although most studies concluded that GM foods do not differ in nutrition or cause any detectable toxic effects in animals, some studies did report adverse changes at a cellular level caused by some GM foods, concluding that "More scientific effort and investigation is needed to ensure that consumption of GM foods is not likely to provoke any form of health problem". A review published in 2009 by Dona and Arvanitoyannis concluded that "results of most studies with GM foods indicate that they may cause some common toxic effects such as hepatic, pancreatic, renal, or reproductive effects and may alter the hematological, biochemical, and immunologic parameters".

However responses to this review in 2009 and 2010 note that the Dona and Arvanitoyannis concentrated on articles with an anti-GM bias that have been refuted by scientists in peer-reviewed articles elsewhere - for example the 35S promoter, stability of transgenes, antibiotic marker genes and the claims for toxic effects of GM foods. In 2007, a review by Domingo of the toxicity by searching in the Publimed data base using 12 search terms, cited 68 references, found that the "number of references" on the safety of GM/transgenic crops was "surprisingly limited" and questioned whether the safety of genetically modified food has been demonstrated; the review also remarked that its conclusions were in agreement with three earlier reviews by Zdunczyk (2001), Bakshi (2003), and Pryme and Lembcke (2003). However, an article in 2007 by Vain found 692 research studies focusing on GM crop and food safety and identified a strong increase in the publication of such articles in recent years. Vain commented that the multidisciplinarian nature of GM research complicates the retrieval of GM studies and requires using many search terms (he used more than 300) and multiple data bases.

Safety Assessments

The starting point for the safety assessment of genetically engineered food products is to assess if the food is "substantially equivalent" to its natural counterpart. To decide if a modified product is substantially equivalent, the product is tested by the manufacturer for unexpected changes in a limited set of components such as toxins, nutrients or allergens that are present in the unmodified food. If these tests show no significant difference between the modified and unmodified products, then no further food safety testing is required. The manufacturers data is then assessed by an independent regulatory body, such as the U.S. Food and Drug Administration.

However, if the product has no natural equivalent, or shows significant differences from the unmodified food, then further safety testing is carried out. A 2003 review in *Trends in Biotechnology* identified 7 main parts of a standard safety test:

1. Study of the introduced DNA and the new proteins or metabolites that it produces;
2. Analysis of the chemical composition of the relevant plant parts, measuring nutrients, anti-nutrients as well as any natural toxins or known allergens;
3. Assess the risk of gene transfer from the food to microorganisms in the human gut;
4. Study the possibility that any new components in the food might be allergens;
5. Estimate how much of a normal diet the food will make up;
6. Estimate any toxicological or nutritional problems revealed by this data;
7. Additional animal toxicity tests if there is the possibility that the food might pose a risk.

This process was examined further in a review published by Kuiper *et al.* 2002 in the journal *Toxicology*, which stated that substantial equivalence does not itself measure risks, but instead identifies differences between existing products and new foods, which might pose dangers to health. If differences do exist, identifying these differences is a starting point for a full safety assessment, rather than an end point. The authors concluded that "The concept of substantial equivalence is an adequate tool in order to identify safety issues related to genetically modified products that have a traditional counterpart". However, the review also noted difficulties in applying this standard in practice, including the fact that traditional foods contain many chemicals that have toxic or carcinogenic effects and that our existing diets therefore have not been proven to be safe. This lack of knowledge on unmodified food poses a problem, as GM foods may have differences in anti-nutrients and natural toxins that have never been identified in the original plant, raising the possibility that harmful changes could be missed.

The application of substantial equivalence has also been more strongly criticized. For example, in a speech in 1999, Andrew Chesson of the University of Aberdeen, stated that substantial equivalence testing "could be flawed in some cases" and that some current safety tests could allow harmful substances to enter the human food chain. In a commentary in *Nature* Millstone *et al.* argued that all GM foods should have extensive biological, toxicological and immunological tests and that the concept of substantial equivalence based solely on chemical analyzes of the components of a food should be abandoned. They stated that this is necessary since it is currently impossible to predict the biological properties of a substance only from knowledge of its chemistry. This commentary was controversial and was criticized for misleading presentation of data and presenting an over-simplified version of safety assessments. For example, Kuiper *et al.* responded to this criticism by noting that equivalency testing does involve more than chemical tests and may include toxicity testing.

Medical writer Barbara Keeler and Marc Lappé argued in a 2001 article in the Los Angeles Times that the differences between Genetically Modified and conventional foods challenges the presumption of equivalence. Using Roundup ready soy that has been on the market since 1995 as an example, they noted the differences when compared to its unmodified counterpart. Significantly lower levels of protein than unmodified soy. Significantly lower levels of phenylalanine, an essential amino acid and as a dietary supplement, the reason doctors advise the consumption of soy products. Levels of trypsin inhibitor were 27% higher and after toasting, Lectin was double that found in conventional soy, both are known allergens. GM soy also has 29% less choline, a B-complex vitamin. Round up ready soy had also stunted the growth of rats in Monsanto's study but had not effected cattle although it had increased the fat content of their milk. The authors do not maintain that modified soy is a hazard but that the FDA accepting such significant differences as being *substantially equivalent* illustrates the need for more rigorous testing, and preferably not by the biotech industries themselves.

The value of current independent studies is problematic as, due to restrictive End-user agreements, researchers are forbidden by law from publishing independent research in peer reviewed journals without the approval of the agritech companies. Cornell University's Elson Shields, the spokesperson for a group of scientists who oppose this practice, submitted a statement to the United States Environmental Protection Agency (EPA) protesting that "as a result of restrictive access, no truly independent research can be legally conducted on many critical questions regarding the technology". Scientific American noted that several studies that were initially approved by seed companies were later blocked from publication when they returned "unflattering" results. While recognising that seed companies' intellectual property rights need to be protected, Scientific American calls the practice dangerous and has called for the restrictions on research in the End-user agreements to be lifted immediately and for the EPA to require, as a condition of approval, that independent researchers have unfettered access to GM products for testing.

The Welsh pressure group GM Free Cymru argues that governments should use independent studies rather than industry studies to assess crop safety. GM Free Cymru has also stated that independently funded researcher, Professor Bela Darvas of Debrecen University was refused Mon 863 Bt corn to use in his studies after previously publishing that a different variety of Monsanto corn was lethal to two Hungarian protected insect species and an insect classified as a rare.

Allergenicity

Worldwide, reports of allergies to all kinds of foods, particularly nuts, fish and shellfish, seem to be increasing, but it is not known if this reflects a genuine change in the risk of allergy, or an increased awareness of food allergies by the public. Some environmental organizations, such as the European Green Party and Greenpeace, have suggested that GM food might trigger food allergies, although other environmentalists have implicated causes as diverse as the greenhouse effect increasing pollen levels, greater exposure to synthetic chemicals, cleaner lifestyles, or more mold in buildings. A 2005 review in the journal *Allergy* of the results from allergen testing of current GM foods stated that "no biotech proteins in foods have been documented to cause allergic reactions".

A well-known case of a GM plant that did not reach the market due to it producing an allergic reaction was a new form of soybean intended for animal feed. The allergen was transferred unintentionally from the Brazil nut into genetically engineered soybeans, in a bid to improve soybean nutritional quality for animal feed use. This new protein increased the levels in the GM soybean of the natural essential amino acid methionine, which is commonly added to poultry feed. Investigation of the GM soybeans revealed that they produced immune reactions in people with Brazil nut allergies, since the methionine rich protein chosen by Pioneer Hi-Bred happened to be a major source of Brazil nut allergy. Although this soybean strain was not developed as a human food, Pioneer Hi-Bred discontinued further development of the GM soybean, due to the difficulty in ensuring that none of these soybeans entered the human food chain.

In November 2005 a pest-resistant field pea developed by the Australian CSIRO for use as a pasture crop was shown to cause an allergic reaction in mice. Work on this variety was immediately halted. The protein added to the pea did not cause the reaction in humans or mice in isolation, but when it was expressed in the pea, it exhibited a subtly different structure which may have caused the allergic reaction. The immunologist who tested the pea noted that crops need to be evaluated case-by-case, and a Greenpeace activist commented that to his knowledge, no countries required feeding studies for approval of genetically modified foods.

Plant scientist Maarten J Chrispeels has made these comments about this example: The recent Prescott et al. paper in JFAC contains a very interesting study on the immunogenicity of amylase [starch digestion enzyme] inhibitor in its native form (isolated from beans) and expressed as a transgene in peas. First of all, amylase inhibitor is a food protein, but also a "toxic" protein because it inhibits our digestive amylases. This is one of the reasons you have to cook your beans! (The other toxic bean protein is phytohemagglutinin and it is much more toxic).

This particular amylase inhibitor is found in the common bean (other species have other amylase inhibitors). Even though it is a food protein, it is unlikely ever to be used for genetic engineering of human foods because it inhibits our amylases. What the results show is that the protein, when synthesized in pea cotyledons has a different immunogenicity than when it is isolated from bean cotyledons (the native form). This is somewhat surprising but may be related to the presence of slightly different carbohydrate chains.

These cases of products that failed safety testing can either be viewed as evidence that genetic modification can produce unexpected and dangerous changes in foods, or alternatively that the current tests are effective at identifying any safety problems before foods come on the market.

Genetic modification can also be used to remove allergens from foods, which may, for example, allow the production of soy products that would pose a smaller risk of food allergies than standard soybeans. A hypo-allergenic strain of soybean was tested in 2003 and shown to lack the major allergen that is found in the beans. A similar approach has been tried in ryegrass, which produces pollen that is a major cause of hayfever: here a fertile GM grass was produced that lacked the main pollen allergen, demonstrating that the production of hypoallergenic grass is also possible.

Environmental Risks and Benefits

The large scale growth of GM plants may have both positive and negative effects on the environment. These may be both direct effects, on organisms that feed on or interact with the crops, or wider effects on food chains produced by increases or decreases in the numbers of other organisms. As an example of benefits, insect-resistant Bt-expressing crops will reduce the number of pest insects feeding on these plants, but as there are fewer pests, farmers do not have to apply as much insecticide, which in turn tends to increase the number of non-pest insects in these fields. A 2006 study of the global impact of GM crops, published by the UK consultancy PG Economics and funded by Illinois-Missouri Biotechnology Alliance, concluded that globally, the technology reduced pesticide spraying by 286,000 tons in 2006, decreasing the environmental impact of herbicides and pesticides by 15%. By reducing the amount of ploughing needed, GM technology led to reductions of greenhouse gases from soil equivalent to removing 6.56 million cars from the roads.

However, a 2009 study published by the Organic Center stated that the use of genetically engineered corn, soybean, and cotton increased the use of herbicides by 383 million pounds (191,500 tons), and pesticide use by 318.4 million pounds (159,200 tons). As an example of a concern about environmental risk, a lab at Cornell University published an article which caused worry in the US that Bt-corn pollen might affect the monarch butterfly. However this concern was disproved by six comprehensive articles in the Proceedings of the National Academy of Sciences in 2001. Monarch populations increased, despite increased Bt corn probably due to reduced pesticide use. Other possible effects might come from the spread of genes from modified plants to unmodified relatives, which might produce species of weeds resistant to herbicides. In some areas of the US "superweeds" have evolved naturally, these weeds are resistant to herbicides and have forced farmers to return to traditional crop management practices.

There has been controversy over the results of a farm-scale trial in the United Kingdom comparing the impact of GM crops and conventional crops on farmland biodiversity. Some claimed that the results showed that GM crops had a significant negative impact on wildlife. They pointed out that the studies showed that using herbicide resistant GM crops allowed better weed control and that under such conditions there were fewer weeds and fewer weed seeds. This result was then extrapolated to suggest that GM crops would have significant impact on the wildlife that might rely on farm weeds. The President of the Royal Society, the body that had carried out the trials, stated that "To generalize and declare 'all GM is bad' or 'all GM is good' for the environment as a result of these experiments is a gross over-simplification", arguing that although the trials showed that the combination of some GM crops with long-lasting herbicides were bad for biodiversity, using other GM crops without these herbicides increased biodiversity.

In July 2005 British scientists showed that transfer of a herbicide-resistance gene from GM oilseed rape to a wild cousin, charlock, and wild turnips was possible. Many agricultural scientists and food policy specialists view GM crops as an important element in sustainable food security and environmental management. This point of view is summarized in the ABIC Manifesto:

On our planet, 18% of the land mass is used for agricultural production. This fraction cannot be increased substantially. It is absolutely essential that the yield per unit of land increases beyond current levels given that: The human population is still growing, and will reach about nine billion by 2040;70,000 km^2 of agricultural land (equivalent to 60% of the German agricultural area) are lost annually to growth of cities and other non-agricultural uses; Consumer diets in developing countries are increasingly changing from plant-based proteins to animal protein, a trend that requires a greater amount of crop-based feeds.

Other scientists, such as Dr. Charles Benbrook, argue that improvement of global food security is hardly being addressed by genetic research and that a lack of yield is often not caused by insufficient genetic resources. Regarding the issues of intellectual property and patent law, an international report from the year 2000 states:

If the rights to these tools are strongly and universally enforced - and not extensively licensed or provided pro bono in the developing world - then the potential applications of GM technologies described previously are unlikely to benefit the less developed nations of the world for a long time (ie until after the restrictions conveyed by these rights have expired).

Issues with Bt Maize

A well publicized claim associated with Bt crops (or transgenic maize) was the concern that pollen from Bt maize might kill the monarch butterfly. This report was puzzling because the pollen from most maize hybrids contains much lower levels of Bt than the rest of the plant and led to multiple follow-up studies. One possible issue revealed in these studies is the possibility that the initial study was flawed; based on the way the pollen was collected, in that they collected and fed non-toxic pollen that was mixed with anther walls that did contain Bt toxin. A collaborative research exercise was carried out over two years by several groups of scientists in the US and Canada, looking at the effects of Bt pollen in both the field and the laboratory.

This resulted in a risk assessment that concluded that any risk posed by the corn to butterfly populations under real-world conditions was negligible. The USDA has stated that the weight of the evidence is that Bt crops do not pose a risk to the monarch butterfly. An independent 2002 review of the scientific literature concluded that "the commercial large-scale cultivation of current Bt–maize hybrids did not pose a significant risk to the monarch population" and noted that despite large-scale planting of GM crops, the butterfly's population is increasing.

In 2007 Andreas Lang, Éva Lauber and Béla Darvas criticized these studies, arguing that there can be a great difference in the effects between the acute exposure tested for and chronic exposure. Moreover, they stated that the "worst case conditions" performed were not in fact worst case scenarios, as laboratory conditions with ample food supply and a favorable climate ensure healthy subjects. They instead believe that in the wild, low temperatures, rain and parasites and disease might exacerbate a Bt effect on butterfly larvae. Their own experiments suggested that some butterfly species were negatively affected by such chronic exposure. Jörg Romeis, who conducted the original studies, replied that if species of Butterfly are affected as Darvas claims that a "more comprehensive assessment will be needed and, depending on the degree and nature of concern, this may extend to field testing".

A 2001 report in *Nature* presented evidence that Bt maize was cross-breeding with unmodified maize in Mexico, although the data in this paper was later described as originating from an artifact and *Nature* stated that "the evidence available is not sufficient to justify the publication of the original paper". A subsequent large-scale study, in 2005, failed to find any evidence of contamination in Oaxaca. However, other authors have stated that they also found evidence of cross-breeding between natural maize and transgenic maize.

There is also a risk that for example, transgenic maize will crossbreed with wild grass variants, and that the Bt-gene will end up in a natural environment, retaining its toxicity. An event like this would have ecological implications. However, there is no evidence of crossbreeding between maize and wild grasses.

In 2009 it was reported that 82,000 hectares (200,000 acres) of Bt corn in South Africa failed to produce seeds. Monsanto claimed average yield was reduced by 25% in those fields affected, it compensated the farmers concerned and the corn varieties were affected by a mistake made in the seed breeding process. Marian Mayet an environmental activitist and director of the Africa-centre for biosecurity in Johannesburg called for a government investigation and asserted that the biotechnology was at fault, "*You cannot make a 'mistake' with three different varieties of corn*". In 2009 South African farmers planted 1,900,000 hectares (4,700,000 acres) of GM maize (73% of the total crop).

As of 2007, a phenomenon called Colony Collapse Disorder (CCD) was noticed in bee hives all over North America, and elsewhere. Although it is not certain if this is a new phenomenon, initial ideas on the possible causes ranged from poor nutrition, infections, parasites and pesticide use. More unusual speculations included radio waves from cellphone base stations, climate change, and the use of transgenic crops containing Bt. The Mid-Atlantic Apiculture Research and Extension Consortium published a report on 2007-03-27 that found no evidence that pollen from Bt crops is adversely affecting bees. Several researchers in the US have since attributed CCD to the spread of a new virus called Israeli acute paralysis virus, although other parasites have also been implicated.

Legal Issues in the US

Alfalfa

On 21 June 2010, the US Supreme Court in *Monsanto v. Geertson Seed Farms* issued its first ruling about GM crops. This case relates to Roundup Ready alfalfa and dates back to 2006, when a coalition of groups led by the Center for Food Safety raised concerns about environmental impacts that the United States Department of Agriculture (USDA) allegedly failed to address before approving the planting of Roundup Ready alfalfa. In response, in 2007, Judge Charles Breyer of the California Northern District Court ruled that USDA was in error when it approved the planting of Roundup Ready alfalfa.

In June 2009, a divided three-judge panel on the 9th U.S. Circuit Court of Appeals upheld Judge Charles Breyer's decision, and Monsanto appealed to the Supreme Court. The impact of the current US Supreme Court ruling is somewhat unclear, with both sides appearing to claim victory.

Sugar Beets

There is also a somewhat similar case involving sugar beets before the same California Northern District Court. This is a case involving Monsanto's breed of pesticide-resistant sugar beets. While Judge Jeffrey S. White (issuing his ruling in the spring of 2010) allowed the planting of GM sugar beets to continue, he also warned that this may be blocked in the future while an environmental review is taking place. Finally, on 13 August 2010, Judge Jeffrey S. White ordered the halt to the planting of the genetically modified sugar beets in the US. He indicated that "the Agriculture Department had not adequately assessed the environmental consequences before approving them for commercial cultivation."

In 2010, before the ruling, 95% of the sugar beet grown in the US was GM. About half the sugar supply in the US came from sugar beet.

Control of the Market

Patent holder companies use their control of their own GMO to corner the market and gain profit. However, other companies often compete for the little market share available to GM foods worldwide. Detractors such as Greenpeace say that patent rights give corporations a dangerous amount of control over their product while corporations claim that they need product control in order to prevent seed piracy, fulfill financial obligations to shareholders and invest in further GM development. Governments have also sought to protect their commercial interests through punitive measures against countries resisting GM foods on ethical or scientific grounds: after moves in France to ban a Monsanto GM corn variety, the US embassy recommended that 'we calibrate a target retaliation list that causes some pain across the EU'.

Controversial Cases

Pusztai Potato Controversy

A large media event occurred in 1998 when scientist Árpád Pusztai, who was considered by some to be a world expert on plant lectins, reported in a television interview that he had found that rats fed potatoes genetically modified by the English biotech company *Cambridge Agricultural Genetics* to contain lectin, a natural insecticide in snowdrop plants, had stunted growth and immune system damage. Initially the Rowett Institute's director, Philip James, congratulated Pusztai after the interview, but his attitude changed after confusion arose over whether the lectin was from a jackbean or a snowdrop (the jackbean lectin is poisonous to the intestines). A press release from the Rowett Institute stated that the lectin had come from a jackbean, but no experiments were made using lectins from the jackbean; in the confusion, it was alleged that Pusztai hadn't actually carried out the experiments. James suspended Pusztai and used misconduct procedures to seize the data. His annual contract was not renewed and both Pusztai and his wife were banned from speaking publicly. In October 1998 the Rowett Institute published an audit criticizing Pusztai's results, which, along with Pusztai's raw data, was sent to six anonymous reviewers who criticized Pusztai's work. Pusztai responded that the raw data was "never intended for publication under intense scrutiny". Pusztai sent the audit report and his rebuttal to scientists who requested it, and in February 1999, twenty-one European and American scientists, mostly friends or acquaintances of Pusztai, orchestrated by the environmental group Friends of the Earth, released a memo supporting Pusztai. Stanley Ewen, who worked with Pusztai, conducted a followup study supporting Pusztai's work and presented the work to a lectin meeting in Sweden.

In October 1999 Pusztai's research was published in the journal The Lancet. Because of the controversial nature of his research the data in this paper, co-authored by Stanley Ewen, was seen by a total of six reviewers when presented for peer review; five of these reviewers judged the work acceptable, although one of the five "deemed the study flawed but favored publication to avoid suspicions of a conspiracy against Pusztai and to give colleagues a chance to see the data for themselves".

The paper did not mention stunted growth or immunity issues, but reported that rats fed on potatoes genetically modified with the snowdrop lectin had "thickening in the mucosal lining of their colon and their jejunum" when compared with rats fed on non modified potatoes. Three Dutch scientists criticized the study on the grounds that the unmodified potatoes were not a fair control diet, and that any rats fed only on potatoes will suffer from a protein deficiency; Pusztai responded to these criticisms by stating that the protein and energy were comparable, and that "a sample size of six is perfectly normal in studies like this".

The genetically modified potatoes were later destroyed by *Cambridge Agricultural Genetics* as were all details of their modification.

Bioequivalence Study of a Corn Cultivar

A controversy arose around biotech company Monsanto's data on a 90-Day Rat Feeding Study on the MON863 strain of GM corn. In May 2005, critics of GM foods pointed to differences in kidney size and blood composition found in this study, suggesting that the observed differences raises questions about the regulatory concept of substantial equivalence. Anti-GM campaigner Jeffrey M. Smith, writing in Biophile Magazine, quoting comments from Pusztai and Seralini,has stated that nutritional studies typically use young, fast-growing animals with starting weights not varying by more than 2% from the average whereas Monsanto's research design used a mix of young and old animals with starting weights ranging from 198.4 to 259.8 grams. Seralini and two other authors published a study of these data, funded by Greenpeace, in 2007 making similar points. The raising of this issue prompted the European Food Safety Authority (EFSA) to reexamine the safety data on this strain of corn.

The EFSA concluded that the observed small numerical decrease in rat kidney weights were not biologically meaningful, and the weights were well within the normal range of kidney weights for control animals. There were no corresponding microscopic findings in the relevant organ systems, and they stated that all blood chemistry and organ weight values fell within the "normal range of historical control values" for rats. In addition the EFSA review stated that the statistical methods used by Séralini et al. in the analysis of the data were incorrect. The European Commission has approved the ÌÏÍ863 corn for animal and human consumption. Food Standards Australia New Zealand reviewed the 2007 Seralini et al study and concluded that "...all of the statistical differences between rats fed MON 863 corn and control rats are attributable to normal biological variation."

Greenpeace stated in a 2007 press release that Séralini et al. had completed a similar analysis of the NK603 strain of corn and came to similar conclusions as they did in their previous study. Séralini et al included this in a re-analysis of three existing rat feeding studies published in 2009.

The European Food Safety Authority reviewed the 2009 Seralini et al paper and concluded that the author's claims were not supported by the data in their paper. They noted that many of their fundamental statistical criticisms of the 2007 paper also applied to the 2009 paper. There was no new information that would change the EFSA's conclusions that the three GM maize types were safe for human, animal health and the environment The French High Council of Biotechnologies Scientific Committee (HCB) also reviewed the 2009 study and concluded that it "..presents no admissible scientific element likely to ascribe any haematological, hepatic or renal toxicity to the three re-analysed GMOs." The HCB also questioned the author's independence. Food Standards Australia New Zealand concluded that the results from the 2009 Séralini et al. study were due to chance alone.

Contamination Issues

In the 1990s genetically modified Flax tolerant to herbicide residues in soil was developed by the Crop Development Centre (CDC) at the University of Saskatchewan in Canada. Named Flax variety FP967, but commonly called *CDC Triffid*, research was controversially halted following protests from Canadian farmers who stood to lose up to 70% of their traditional export markets if it was introduced. GM Flax was deregistered, its sale was criminalized and in 2001 all modified seeds were destroyed. No modified crops had been planted and no seed had been sold but GM industry proponent Alan McHughen controversially passed out sample packets of seeds at presentations. In early September 2009, Flax imported into Germany was found to be contaminated with *CDC Triffid* causing the price of Canadian Flax to fall 32 percent. By mid November 35 countries reported contamination of imported Canadian Flax which has now been banned by the European Union. Canadian farmers are expected to be responsible for the cost of the cleanup and testing of future crops.

In 2000, Aventis StarLink corn, which had been approved only as animal feed due to concerns about possible allergic reactions in humans, was found contaminating corn products in U.S. supermarkets. An episode involving Taco Bell taco shells was particularly well publicized which resulted in sales of StarLink seed being discontinued. The registration for the Starlink varieties was voluntarily withdrawn by Aventis in October 2000. Aid sent by the UN and the US to Central African nations was also found to be contaminated with StarLink corn and the aid was rejected. The US corn supply has been monitored for Starlink Bt proteins since 2001 and no positive samples have been found since 2004.

Public Perception

Research by the Pew Initiative on Food and Biotechnology has shown that in 2005 Americans' knowledge of genetically modified foods and animals continues to remain low, and their opinions reflect that they are particularly uncomfortable with animal cloning. In one instance of consumer confusion, DNA Plant Technology's Fish tomato transgenic organism was conflated with Calgene's Flavr Savr transgenic food product. The Pew survey also showed that despite continuing concerns about GM foods, American consumers do not support banning new uses of the technology, but rather seek an active role from regulators to ensure that new products are safe.

Only 2% of Britons were said to be "happy to eat GM foods", and more than half of Britons were against GM foods being available to the public, according to a 2003 study. However a 2009 review article of European consumer polls concluded that opposition to GMOs in Europe has been gradually decreasing. Approximately half of European consumers accepted gene technology, particularly when benefits for consumers and for the environment could be linked to GMO products. 80 % of respondents did not cite the application of GMOs in agriculture as a significant environmental problem. Many consumers seem unafraid of health risks from GMO products and most European consumers did not actively avoid GMO products while shopping.

In Australia, GM foods that have novel DNA, novel protein, altered characteristics or has to be cooked or prepared in a different way compared to the conventional food have, since December 2001, had to be identified on food labels. However, multiple surveys have shown that while 45% of the public will accept GM foods, some 93% demand all genetically modified foods be labelled as such. A 2007 survey by the Food Standards Australia and New Zealand found that 27% of Australians looked at the label to see if it contained GM material when purchasing a grocery product for the first time. Labelling legislation has been introduced and rejected several times since 1996 on the grounds of "restraint of trade" due to the cost of labelling. The controversy erupted again in 2009 when Graincorp, the nations largest grain handler, announced it would mix GM Canola with its unmodified grain. Traditional growers, who largely rely on GM-free markets, had been told they would need to pay to have their produce certified GM free. Graincorp reversed its decision the same year. Critics such as Greenpeace and the Gene Ethics Network have renewed calls for more labelling.

Opponents of genetically modified food often refer to it as "Frankenfood", after Mary Shelley's character Frankenstein and the monster he creates, in her novel of the same name. The term was coined in 1992 by Paul Lewis, an English professor at Boston College who used the word in a letter he wrote to the *New York Times* in response to the decision of the US Food and Drug Administration to allow companies to market genetically modified food. The term "Frankenfood" has become a battle cry of the European side in the US-EU agricultural trade war.

Critics have protested in regards to the appointment of pro GM lobbyists to senior positions in the FDA. Michael R. Taylor has been appointed as a senior adviser to the FDA on food safety and Dennis Wolff is expected to take up the position of Under-Secretary of the newly created Agriculture for Food Safety. Taylor is a former Monsanto lobbyist credited as being responsible for the implementation of "substantial equivalence" in place of food safety studies and for his advocacy that resulted in the Delaney clause that prohibited the inclusion of "any chemical additive found to induce cancer in man.. or animals" in processed foods being amended in 1996 to allow the inclusion of pesticides in GMOs. Wolff is the Pennsylvania Secretary of Agriculture who successfully lobbied to ban organic farmers from labelling their products as being GM free and was a proponent of the "ACRE" initiative which gave the Pennsylvania state attorney general's office the authority to sue municipalities that banned GMOs. Several anti-GMO organisations have organised petitions demanding Taylor's resignation and opposing Wolff's appointment and also conducted letter writing campaigns protesting the conflict of interest.

Genetically Modified Organism

A genetically modified organism (GMO) or genetically engineered organism (GEO) is an organism whose genetic material has been altered using genetic engineering techniques. These techniques, generally known as recombinant DNA technology, use DNA molecules from different sources, which are combined into one molecule to create a new set of genes. This DNA is then transferred into an organism, giving it modified or novel genes. Transgenic organisms, a subset of GMOs, are organisms which have inserted DNA that originated in a different species.

Production

Genetic modification involves the insertion or deletion of genes. When genes are inserted, they usually come from a different species, which is a form of horizontal gene transfer. In nature this can occur when exogenous DNA penetrates the cell membrane for any reason. To do this artificially may require attaching the genes to a virus or just physically inserting the extra DNA into the nucleus of the intended host with a very small syringe, or with very small particles fired from a gene gun. However, other methods exploit natural forms of gene transfer, such as the ability of *Agrobacterium* to transfer genetic material to plants, or the ability of lentiviruses to transfer genes to animal cells.

History

The general principle of producing a GMO is to add new genetic material into an organism's genome. This is called genetic engineering and was made possible through the discovery of DNA and the creation of the first recombinant bacteria in 1973; an existing bacterium *E. coli* expressing an exogenic Salmonella gene. This led to concerns in the scientific community about potential risks from genetic engineering, which were first discussed in depth at the Asilomar Conference in 1975. One of the main recommendations from this meeting was that government oversight of recombinant DNA research should be established until the technology was deemed safe. Herbert Boyer then founded the first company to use recombinant DNA technology, Genentech, and in 1978 the company announced creation of an *E. coli* strain producing the human protein insulin. In 1986, field tests of bacteria genetically engineered to protect plants from frost damage (ice-minus bacteria) at a small biotechnology company called Advanced Genetic Sciences of Oakland, California, were repeatedly delayed by opponents of biotechnology. In the same year, a proposed field test of a microbe genetically engineered for a pest resistance protein by Monsanto Company was dropped. In the late 1980s and early 1990s guidance on assessing the safety of genetically engineered plants and food emerged from organizations including the FAO and WHO. Small scale experimental plantings of genetically modified (GM) plants began in Canada and the U.S. in the late 1980s. The first approvals for large scale, commercial cultivation came in the mid 1990s. Since that time, adoption of GM plants by farmers has increased annually.

Uses

GMOs are used in biological and medical research, production of pharmaceutical drugs, experimental medicine (e.g. gene therapy), and agriculture (e.g. golden rice). The term "genetically modified organism" does not always imply, but can include, targeted insertions of genes from one species into another. For example, a gene from a jellyfish, encoding a fluorescent protein called GFP, can be physically linked and thus co-expressed with mammalian genes to identify the location of the protein encoded by the GFP-tagged gene in the mammalian cell. Such methods are useful tools for biologists in many areas of research, including those who study the mechanisms of human and other diseases or fundamental biological processes in eukaryotic or prokaryotic cells.

To date the most controversial but also the most widely adopted application of GMO technology is patent-protected food crops which are resistant to commercial herbicides or are able to produce pesticidal proteins from within the plant, or *stacked trait* seeds, which do both. The largest share of the GMO crops planted globally are owned by the US firm Monsanto. In 2007, Monsanto's trait technologies were planted on 246 million acres (1,000,000 km^2) throughout the world, a growth of 13 percent from 2006. However, patents on the first Monsanto products to enter the marketplace will begin to expire in 2014, democratizing Monsanto products. In addition, a 2007 report from the European Joint Research Commission predicts that by 2015, more than 40 per cent of new GM plants entering the global marketplace will have been developed in Asia.

In the corn market, Monsanto's triple-stack corn—which combines Roundup Ready 2 weed control technology with YieldGard Corn Borer and YieldGard Rootworm insect control—is the market leader in the United States. U.S. corn farmers planted more than 32 million acres (130,000 km^2) of triple-stack corn in 2008, and it is estimated the product could be planted on 56 million acres (230,000 km^2) in 2014–2015. In the cotton market, Bollgard II with Roundup Ready Flex was planted on approximately 5 million acres (20,000 km^2) of U.S. cotton in 2008.

According to the International Service for the Acquisition of Agri-Biotech Applications (ISAAA), of the approximately 14 million farmers who grew biotech crops in 2009, some 90% were resource-poor farmers in developing countries. These include some 7 million farmers in the cotton-growing areas of China, an estimated 5.6 million small farmers in India (Bt cotton), 250,000 in the Philippines, South Africa (biotech cotton, maize and soybeans often grown by subsistence women farmers) and the other twelve developing countries which grew biotech crops in 2009. 10 million more small and resource-poor farmers may have been secondary beneficiaries of Bt cotton in China. The global commercial value of biotech crops grown in 2008 was estimated to be US$130 billion.

In the United States, the United States Department of Agriculture (USDA) reports on the total area of GMO varieties planted. According to National Agricultural Statistics Service, the states published in these tables represent 81–86 percent of all corn planted area, 88–90 percent of all soybean planted area, and 81–93 percent of all upland cotton planted area (depending on the year). USDA does not collect data for global area. Estimates are produced by the International Service for the Acquisition of Agri-biotech Applications (ISAAA) and can be found in the report, "Global Status of Commercialized Transgenic Crops: 2007".

Transgenic animals are also becoming useful commercially. On February 6, 2009 the U.S. Food and Drug Administration approved the first human biological drug produced from such an animal, a goat. The drug, ATryn, is an anticoagulant which reduces the probability of blood clots during surgery or childbirth. It is extracted from the goat's milk.

Detection

Testing on GMOs in food and feed is routinely done by molecular techniques like DNA microarrays or qPCR. The test can be based on screening elements (like p35S, tNos, pat, or bar) or event-specific markers for the official GMOs (like Mon810, Bt11, or GT73). The array-based method combines multiplex PCR and array technology to screen samples for different potential GMOs, combining different approaches (screening elements, plant-specific markers, and event-specific markers). The qPCR is used to detect specific GMO events by usage of specific primers for screening elements or event-specific markers.

To avoid any kind of false positive or false negative testing outcome, comprehensive controls for every step of the process is mandatory. A CaMV check is important to avoid false positive outcomes based on virus contamination of the sample.

Transgenic Microbes

Bacteria were the first organisms to be modified in the laboratory, due to their simple genetics. These organisms are now used for several purposes, and are particularly important in producing large amounts of pure human proteins for use in medicine. Genetically modified bacteria are used to produce the protein insulin to treat diabetes. Similar bacteria have been used to produce clotting factors to treat haemophilia, and human growth hormone to treat various forms of dwarfism.

Transgenic Animals

Transgenic animals are used as experimental models to perform phenotypic and for testing in biomedical research. Other applications include the production of human hormones such as insulin.

Fruit Flies

In biological research, transgenic fruit flies (*Drosophila melanogaster*) are model organisms used to study the effects of genetic changes on development. Fruit flies are often preferred over other animals due to their short life cycle, low maintenance requirements, and relatively simple genome compared to many vertebrates.

Mammals

Genetically modified mammals are an important category of genetically modified organisms. Transgenic mice are often used to study cellular and tissue-specific responses to disease.

In 1999, scientists at the University of Guelph in Ontario, Canada created the genetically engineered Enviropig. The Enviropig excretes from 30 to 70.7% less phosphorus in manure depending upon the age and diet. In February 2010, Environment Canada determined that Enviropigs are in compliance with the Canadian Environmental Protection Act and can be produced outside of the research context in controlled facilities where they are segregated from other animals. In 2009, scientists in Japan announced that they had successfully transferred a gene into a primate species (marmosets) and produced a stable line of breeding transgenic primates for the first time.

Cnidarians

Cnidarians such as *Hydra* and the sea anemone Nematostella vectensis have become attractive model organisms to study the evolution of immunity and certain developmental processes. An important technical breakthrough was the development of procedures for generation of stably transgenic hydras and sea anemones by embryo microinjection.

Fish

Genetically modified fish have promoters driving an over-production of "all fish" growth hormone. This resulted in dramatic growth enhancement in several species, including salmonids, carps and tilapias.

Gene Therapy

Gene therapy, uses genetically modified viruses to deliver genes that can cure disease into human cells. Although gene therapy is still relatively new, it has had some successes. It has been used to treat genetic disorders such as severe combined immunodeficiency, and treatments are being developed for a range of other currently incurable diseases, such as cystic fibrosis, sickle cell anemia, and muscular dystrophy. Current gene therapy technology only targets the non-reproductive cells meaning that any changes introduced by the treatment can not be transmitted to the next generation. Gene therapy targeting the reproductive cells—so-called "Germ line Gene Therapy"—is very controversial and is unlikely to be developed in the near future.

Transgenic Plants

Transgenic plants have been engineered to possess several desirable traits, such as resistance to pests, herbicides, or harsh environmental conditions, improved product shelf life, and increased nutritional value. Since the first commercial cultivation of genetically modified plants in 1996, they have been modified to be tolerant to the herbicides glufosinate and glyphosate, to be resistant to virus damage as in Ringspot virus-resistant GM papaya, grown in Hawaii, and to produce the Bt toxin, an insecticide that is non-toxic to mammals. Most GM crops grown today have been modified with "input traits", which provide benefits mainly to farmers. The GM oilseed crops on the market today offer improved oil profiles for processing or healthier edible oils. The GM crops in development offer a wider array of environmental and consumer benefits such as nutritional enhancement, drought and stress tolerance.

GM plants are being developed by both private companies and public research institutions such as CIMMYT, the International Maize and Wheat Improvement Centre. Other examples include a genetically modified sweet potato, enhanced with protein and other nutrients, while golden rice, developed by the International Rice Research Institute (IRRI), has been discussed as a possible cure for Vitamin A deficiency.

The coexistence of GM plants with conventional and organic crops has raised significant concern in many European countries. Due to relatively high demand from European consumers for the freedom of choice between GM and non-GM foods, EU regulations require measures to avoid mixing of foods and feed produced from GM crops and conventional or organic crops. European research programs such as Co-Extra, Transcontainer, and SIGMEA are investigating appropriate tools and rules. At the field level, biological containment methods include isolation distance and pollen barriers. Such measures are generally not used in North America because they are very costly and there are no safety-related reasons to employ them.

Cisgenic Plants

Cisgenesis, sometimes also called Intragenesis, is a product designation for a category of genetically engineered plants. A variety of classification schemes have been proposed, that order genetically modified organisms based on the nature of introduced genotypical changes rather than the process of genetic engineering.

While some genetically modified plants are developed by the introduction of a gene originating from distant, sexually incompatible species into the host genome, cisgenic plants contain genes which have been isolated either directly from the host species or from sexually compatible species. The new genes are introduced using recombinant DNA methods and gene transfer. Some scientists hope that the approval process of cisgenic plants might be simpler than that of proper transgenics, but it remains to be seen.

Controversy

Biological Process

The use of genetically modified organisms has sparked significant controversy in many areas. Some groups or individuals see the generation and use of GMO as intolerable meddling with biological states or processes that have naturally evolved over long periods of time, while others are concerned about the limitations of modern science to fully comprehend all of the potential negative ramifications of genetic manipulation. Other people see this as a continuation in the role humanity has occupied for thousands of years, modifying the genetics of crops.

Foodchain

The safety of GMOs in the foodchain has been questioned by some environmental groups, with concerns such as the possibilities that GMOs could introduce new allergens into foods, or contribute to the spread of antibiotic resistance. All studies published to date have shown no adverse health effects resulting from humans eating genetically modified foods, environmental groups still discourage consumption in many countries, claiming that GM foods are unnatural and therefore unsafe. Such concerns have led to the adoption of laws and regulations that require safety testing of any new organism produced for human consumption.

GMOs' proponents note that because of the safety testing requirements imposed on GM foods, the risk of introducing a plant variety with a new allergene or toxin using genetic modification is much smaller than using traditional breeding processes. An example of an allergenic plant created using traditional breeding is the kiwi. One article calculated that the marketing of GM salmon could reduce the cost of salmon by half, thus increasing salmon consumption and preventing 1,400 deaths from heart attack a year in the United States.

Trade in Europe and Africa

In response to negative public opinion, Monsanto announced its decision to remove their seed cereal business from Europe, and environmentalists crashed a World Trade Organization conference in Cancun that promoted GM foods and was sponsored by Committee for a Constructive Tomorrow (CFACT). Some African nations have refused emergency food aid from developed countries, fearing that the food is unsafe. During a conference in the Ethiopian capital of Addis Ababa, Kingsley Amoako, Executive Secretary of the United Nations Economic Commission for Africa (UNECA), encouraged African nations to accept genetically modified food and expressed dissatisfaction in the public's negative opinion of biotechnology.

Agricultural Surpluses

Patrick Mulvany, Chairman of the UK Food Group, accused some governments, especially the Bush administration, of using GM food aid as a way to dispose of unwanted agricultural surpluses. The UN blamed food companies and accused them of violating human rights, calling on governments to regulate these profit-driven firms. It is widely believed that the acceptance of biotechnology and genetically modified foods will also benefit rich research companies and could possibly benefit them more than consumers in underdeveloped nations.

Labelling

While some groups advocate the complete prohibition of GMOs, others call for mandatory labelling of genetically modified food or other products. Other controversies include the definition of patent and property pertaining to products of genetic engineering. According to the documentary Food, Inc. efforts to introduce labelling of GMOs has repetedly met resistance from lobbyists and politicians affiliated with companies like Monsanto.

Testing

Bruce Stutz's article, "Wanted: GM Seeds for Study," highlights a story of two dozen scientists who spoke out against the research restrictions put forth by companies producing genetically modified (GM) seeds such as DuPont, Monsanto, and Syngenta. In February 2009, after scientists warned the U.S. Environmental protection Agency (EPA) "that industry influence had made independent analyses of transgenic crops impossible," the American Seed Trade Association (ASTA) agreed that they "would allow researchers greater freedom to study the effects of GM food crops." This agreement left many scientists optimistic about the future, but there is little optimism as to whether this agreement has the ability to "alter what has been a research environment rife with obstruction and suspicion."

Impoverished Nations

Some groups believe that impoverished nations will not reap the benefits of biotechnology because they do not have easy access to these developments, cannot afford modern agricultural equipment, and certain aspects of the system revolving around intellectual property rights are unfair to "undeveloped countries". For example, The CGIAR (Consultative Group of International Agricultural Research) is an aid and research organization that has been working to achieve sustainable food security and decrease poverty in undeveloped countries since its formation in 1971. In an evaluation of CGIAR, the World Bank praised its efforts but suggested a shift to genetics research and productivity enhancement.

This plan has several obstacles such as patents, commercial licenses, and the difficulty that third world countries have in accessing the international collection of genetic resources and other intellectual property rights that would educate them about modern technology. The International Treaty on Plant Genetic Resources for Food and Agriculture has attempted to remedy this problem, but results have been inconsistent. As a result, "orphan crops", such as teff, millets, cowpeas, and indigenous plants, are important in the countries where they are grown, but receive little investment.

Private Investments

The development and implementation of policies designed to encourage private investments in research and marketing biotechnology that will meet the needs of poverty-stricken nations, increased research on other problems faced by poor nations, and joint efforts by the public and private sectors to ensure the efficient use of technology developed by industrialized nations have been suggested. In addition, industrialized nations have not tested GM technology on tropical plants, focusing on those that grow in temperate climates, even though undeveloped nations and the people that need the extra food live primarily in tropical climates. Some European scientists are concerned that political factors and ideology prevent unbiased assessment of GM technology in some EU countries, with a negative effect on the whole community.

Transgenic Organisms

Another important controversy is the possibility of unforeseen local and global effects as a result of transgenic organisms proliferating. The basic ethical issues involved in genetic research are discussed in the article on genetic engineering.

Some critics have raised the concern that conventionally-bred crop plants can be cross-pollinated (bred) from the pollen of modified plants. Pollen can be dispersed over large areas by wind, animals, and insects. In 2007, the U.S.

Department of Agriculture fined Scotts Miracle-Gro $500,000 when modified genetic material from creeping bentgrass, a new golf-course grass Scotts had been testing, was found within close relatives of the same genus (*Agrostis*) as well as in native grasses up to 21 km (13 miles) away from the test sites, released when freshly cut grass was blown by the wind.

GM proponents point out that outcrossing, as this process is known, is not new. The same thing happens with any new open-pollinated crop variety—newly introduced traits can potentially cross out into neighboring crop plants of the same species and, in some cases, to closely related wild relatives. Defenders of GM technology point out that each GM crop is assessed on a case-by-case basis to determine if there is any risk associated with the outcrossing of the GM trait into wild plant populations. The fact that a GM plant may outcross with a related wild relative is not, in itself, a risk unless such an occurrence has negative consequences. If, for example, an herbicide resistance trait was to cross into a wild relative of a crop plant it can be predicted that this would not have any consequences except in areas where herbicides are sprayed, such as a farm. In such a setting the farmer can manage this risk by rotating herbicides.

The European Union funds research programs such as Co-Extra, that investigate options and technologies on the coexistence of GM and conventional farming. This also includes research on biological containment strategies and other measures to prevent outcrossing and enable the implementation of coexistence.

If patented genes are outcrossed, even accidentally, to other commercial fields and a person deliberately selects the outcrossed plants for subsequent planting then the patent holder has the right to control the use of those crops. This was supported in Canadian law in the case of Monsanto Canada Inc. v. Schmeiser. However, F2 seed (next generation from the patented uniform crop) is highly variable as the crossed genes segregate. This makes it a much more variable crop and much less reliable for resistance or yield, so deliberate saving of seed is not practiced with patented crops. Instead new seed is bought each year, the farmer choosing a higher cost of seed to generate a much lower cost in insecticides/pesticides or for insurance of reliability and yield of the crop.

"Terminator" and "Traitor"

An often cited controversy is a "Technology Protection" technology dubbed 'Terminator'. This uncommercialized technology would allow the production of first generation crops that would not generate seeds in the second generation because the plants yield sterile seeds. The patent for this so-called "terminator" gene technology is owned by *Delta and Pine Land Company* and the United

States Department of Agriculture. Delta and Pine Land was bought by Monsanto Company in August 2006. Similarly, the hypothetical trait-specific Genetic Use Restriction Technology, also known as 'Traitor' or 'T-gut', requires application of a chemical to genetically modified crops to reactivate engineered traits. This technology is intended both to limit the spread of genetically engineered plants, and to require farmers to pay yearly to reactivate the genetically engineered traits of their crops. Genetic Use Restriction Technology is under development by companies including Monsanto and AstraZeneca.

In addition to the commercial protection of proprietary technology in self-pollinating crops such as soybean (a generally contentious issue), another purpose of the terminator gene is to prevent the escape of genetically modified traits from cross-pollinating crops into wild-type species by sterilizing any resultant hybrids. Some environmentalist groups, while considering outcrossing of GM plants dangerous, felt the technology would prevent re-use of seed by farmers growing such terminator varieties in the developing world and was ostensibly a means to exercise patent claims.

However other environmental groups welcomed the terminator gene as a means of preventing GM crops from mixing with natural crops. Hybrid seeds were commonly used in the developed countries long before the introduction of GM crops. Hybrid seeds cannot be saved, so purchasing new seed every year is already a standard agricultural practice.

There are technologies evolving which contain the transgene by biological means and still can provide fertile seeds using fertility restorer functions. Such methods are being developed by several EU research programs, among them Transcontainer and Co-Extra.

Governmental Support and Opposition

Australia

Several states of Australia had placed bans on planting GM food crops, beginning in 2003. However, in late 2007 the states of New South Wales and Victoria lifted their bans. Western Australia lifted their state's ban in December 2008, while South Australia continues its ban. Tasmania has extended its moratorium until November 2014. The state of Queensland has allowed the growing of GM crops since 1995 and has never had a GM ban.

Canada

Genetically modified crops have been widely adopted in Canada and have been grown since 1995. Nearly all of the canola grown in Canada is GM, as are significant proportions of corn and soybean. The Canadian regulatory system for biotechnology is science-based.

In general, biotechnology is well-accepted by Canadian consumers and farmers, with some exceptions. In 2005, a standing committee of the government of Prince Edward Island (PEI) in Canada assessed a proposal to ban the production of GMOs in the province. The ban was not passed. As of January 2008, the use of genetically modified crops on PEI was rapidly increasing. Mainland Canada is one of the world's largest producers of GM canola.

Japan

As of 2009, Japan has no commercial farming of any kinds of genetically modified food. Consumers have strongly resisted both imports and attempts to grow GMO in the country. Campaigns by consumer groups and environmental groups, such as Consumers Union of Japan and Greenpeace Japan, as well as local campaigns, have been very successful.

In Hokkaido, a special bylaw has made it virtually impossible to grow GMOs, as the No! GMO Campaign collected over 200,000 signatures to oppose GMO farming. Consumers Union of Japan participated together with other Japanese NGOs at the Planet Diversity conference in Bonn, Germany on May 12–16, 2008, a global congress on the future of food and agriculture, with a demonstration to celebrate biodiversity, to oppose GMOs. "We don't only need networks between people, but between people and plants, and people and planet earth," noted Koketsu Michiyo from CUJ.

Cross-pollination has commonly occurred in Japan, as canola seed (rape seed) is imported from Canada. Around ports and the roads to major food oil companies, GE canola has now been found growing wild. Imported canola seeds have been found to be GMO varieties, including the Roundup Ready and Liberty Link types not grown in Japan.

Activists and local groups, as well as the No! GMO Campaign and others, are alarmed that imported GMOs may harm the biodiversity and cause irreversible damage. A report from the Japanese National Institute for Environmental Studies (NIES) confirms that herbicide-resistant genetically engineered canola plants were identified in five of the six Japanese ports where samples were collected.

A number of Japanese groups have been making submissions to Western Australia's Review of the Genetically Modified Crops Free Areas Act 2003. These include the Seikatsu Club Consumers' Cooperative Union and the Consumers Union of Japan. Seikatsu—an umbrella group of 29 Seikatsu Club Consumers' Co-Operatives—and its oil crushers Okamura Oil Mill Ltd and Yonezawa Oil Co. Ltd., all have non-GE canola policies. The groups stopped importing canola from Canada after the introduction of GE canola, when cross-pollination made it impossible to guarantee GE-free canola from Canada.

Pakistan

The government supports the use of hybrid seeds. However, Monsanto once tried to sell their hybrid seeds of such important crops as wheat and rice via the government. Even though yields would have increased, it would have made the Pakistani population dependent on the seeds of one company. The contract was never given.

New Zealand

In New Zealand, no genetically modified food is grown and no medicines containing live genetically-modified organisms have been approved for use. However, medicines manufactured using genetically modified organisms that do not contain live organisms have been approved for sale, and imported foods with genetically modified components are sold.

United States

In 2004, Mendocino County, California became the first county in the United States to ban the production of GMOs. The measure passed with a 57% majority. In California, Trinity and Marin counties have also imposed bans on GM crops, while ordinances to do so were unsuccessful in Butte, Lake, San Luis Obispo, Humboldt, and Sonoma counties. Supervisors in the agriculturally-rich counties of Fresno, Kern, Kings, Solano, Sutter, and Tulare have passed resolutions supporting the practice. In 2007, with reference to US negotiations with the EU on agricultural biotechnology, US diplomatic cables recommended that 'we calibrate a target retaliation list that causes some pain across the EU'.

Zambia

The Zambian government has launched a campaign to educate and increase awareness of the benefits of biotechnology, including genetically modified crops, in order to change negative public opinion.

Other Africa

In 2010, after nine years of talks, the Common Market for Eastern and Southern Africa (COMESA) produced a draft policy on GM technology. This proposed policy was sent to all 19 national governments for consultation in September 2010.

Under the policy, a member country which wants to grow a new GM crop would inform COMESA who would have sufficient scientific expertise to make the decision as to whether the crop was safe for the environment and for humans. At the moment, few countries have the resources to make their own decisions. Once COMESA had made their decision, permission would be granted for the crop to be grown in all 19 member countries. Member countries would retain the power not to grow the crop in their own country if they wanted.

France

The cultivation of Monsanto's MON 810 corn was forbidden in France on February 9, of 2008. It was the only GMO authorized in France. The safeguard measure is taken as far as side effects on human health will be known. In 2010 Marion Guillou, president of the National Institute for Agronomical Research and one of France's top farm researcher, said she can no longer work on developing new GMOs due to widespread distrust and even hostility by European consumers.

Germany

Germany placed a ban on the cultivation and sale of GMO maize in April 2009.

Other European Countries

MON 810 (maize) was the first GMO crop to be cultivated in Europe. The initial lines of maize were approved in 1997 and, by 2009, 76,000 hectares of GM maize were grown in Spain (20% of Spain's maize production). Smaller amounts were produced in the Czech Republic, Slovakia, Portugal, Romania and Poland. However, in addition to France and Germany, other European countries that have placed bans on the cultivation and sale of GMOs include Austria, Hungary, Greece, and Luxembourg. Ireland has also banned GMO cultivation, and has instituted a voluntary label for GMO-free food products. Poland has also tried to institute a ban, with backlash from the European Commission. Bulgaria effectively banned cultivation of genetically modified organisms on March 18, 2010.

On 2 March 2010 a second species of GMO, a potato named Amflora, was approved for cultivation for industrial applications in the EU by the European Commission and was grown in Germany, Sweden and the Czech Republic that year. On 13 July 2010, the European Commission issued a recommendation that in future individual states in the EU should be able to ban the growing of specific GM crops that had been scientifically approved at the EU level. A ban could be justified on cultural, economic or ethical grounds. The EU approval process for imports of GM crops and labelling of GM food products remained in place.

Marker Assisted Selection a Way to Select Suitable Offspring without Using Genetic Engineering

Marker assisted selection or marker aided selection (MAS) is a process whereby a marker (morphological, biochemical or one based on DNA/RNA variation) is used for indirect selection of a genetic determinant or determinants of a trait of interest (i.e. productivity, disease resistance, abiotic stress tolerance, and/or quality). This process is used in plant and animal breeding. Considerable developments in biotechnology have led plant breeders to develop more efficient selection systems to replace traditional phenotypic-pedigree-based selection systems.

Marker assisted selection (MAS) is indirect selection process where a trait of interest is selected not based on the trait itself but on a marker linked to it. For example if MAS is being used to select individuals with a disease, the level of disease is not quantified but rather a marker allele which is linked with disease is used to determine disease presence. The assumption is that linked allele associates with the gene and/or quantitative trait locus (QTL) of interest. MAS can be useful for traits that are difficult to measure, exhibit low heritability, and/or are expressed late in development.

Marker Types

A marker may be:

- Biological- Different pathogen races or insect biotypes based on host pathogen or host parasite interaction can be used as a marker since the genetic constitution of an organism can affect its susceptibility to pathogens or parasites.
- Morphological - First markers loci available that have obvious impact on morphology of plant. Genes that affect form, coloration, male sterility or resistance among others have been analysed in many plant species. Examples of this type of marker may include the presence or absence of awn, leaf sheath coloration, height, grain color, aroma of rice etc. In well-characterized crops like maize, tomato, pea, barley or wheat, tens or even hundreds of such genes have been assigned to different chromosomes.
- Biochemical- A gene that encodes a protein that can be extracted and observed; for example, isozymes and storage proteins.
- Cytological - The chromosomal banding produced by different stains; for example, G banding.
- DNA-based and/or molecular- A unique (DNA sequence), occurring in proximity to the gene or locus of interest, can be identified by a range of molecular techniques such as RFLPs, RAPDs, AFLP, DAF, SCARs, microsatellites etc.

Sax in 1923 first reported association of a simply inherited genetic marker with a quantitative trait in plants when he observed segregation of seed size associated with segregation for a seed coat color marker in beans (*Phaseolus vulgaris* L.). Rasmusson in 1935 demonstrated linkage of flowering time (a quantitative trait) in peas with a simply inherited gene for flower color.

Gene vs Marker

The gene of interest is directly related with production of protein(s) that produce certain phenotypes whereas markers should not influence the trait of interest but are genetically linked (and so go together during segregation of gametes due to the concomitant reduction in homologous recombination between the marker and gene of interest). In many traits genes are discovered and can be directly assayed for their presence with a high level of confidence. However, if a gene is not isolated marker's help is taken to tag a gene of interest. In such case there may be some false positive results due to recombination between marker of interest and gene (or QTL). A perfect marker would elicit no false positive results.

Important Properties of Ideal Markers for MAS

An ideal marker:

- Easy recognition of all possible phenotypes (homo- and heterozygotes) from all different alleles
- Demonstrates measurable differences in expression between trait types and/or gene of interest alleles, early in the development of the organism
- Has no effect on the trait of interest that varies depending on the allele at the marker loci
- Low or null interaction among the markers allowing the use of many at the same time in a segregating population
- Abundant in number
- Polymorphic.

Demerits of Morphological Markers

Morphological markers are associated with several general deficits that reduce their usefulness including:

- the delay of marker expression until late into the development of the organism
- dominance
- deleterious effects
- pleiotropy
- confounding effects of genes unrelated to the gene or trait of interest but which also affect the morphological marker (epistasis)
- rare polymorphism
- frequent confounding effects of environmental factors which affect the morphological characteristics of the organism.

To avoid problems specific to morphological markers, the DNA-based markers have been developed. They are highly polymorphic, simple inheritance (often codomimant), abundantly occur throughout the genome, easy and fast to detect, minimum pleiotropic effect and detection is not dependent on the developmental stage of the organism. Numerous markers have been mapped to different chromosomes in several crops including rice, wheat, maize, soybean and several others. Those markers have been used in diversity analysis, parentage detection, DNA fingerprinting, and prediction of hybrid performance. Molecular markers are useful in indirect selection processes, enabling manual selection of individuals for further propagation.

Selection for Major Genes Linked to Markers

The major genes which are responsible for economically important characteristics are frequent in the Plant Kingdom. Such characteristics include disease resistance, male sterility, self-incompatibility, others related to shape, color, and architecture of whole plants and are often of mono- or oligogenic in nature. The marker loci which are tightly linked to major genes can be used for selection and are sometimes more efficient than direct selection for the target gene. Such vantages in efficiency may be due for example, to higher expression of the marker mRNA in such cases that the marker is actually a gene. Alternatively, in such cases that the target gene of interest differs between two alleles by a difficult-to-detect single nucleotide polymorphism, an external marker (be it another gene or a polymorphism that is easier to detect, such as a short tandem repeat) may present as the most realistic opti.

Situations that are Favorable for Molecular Marker Selection

There are several indications for the use of molecular markers in the selection of a genetic trait.

In such situations that:

- the selected character is expressed late in plant development, like fruit and flower features or adult characters with a juvenile period (so that it is not necessary to wait for the organism to become fully developed before arrangements can be made for propagation)
- the expression of the target gene is recessive (so that individuals which are heterozygous positive for the recessive allele can be crossed to produce some homozygous offspring with the desired trait)
- there is requirement for the presence of special conditions in order to invoke expression of the target gene(s), as in the case of breeding for disease and pest resistance (where inoculation with the disease or subjection to pests would otherwise be required). This advantage derives

from the errors due to unreliable inoculation methods and the fact that field inoculation with the pathogen is not allowed in many areas for safety reasons. Moreover, problems in the recognition of the environmentally unstable genes can be eluded.

- the phenotype is affected by two or more unlinked genes (epistatis). For example, selection for multiple genes which provide resistance against diseases or insect pests for gene pyramiding.

The cost of genotyping (an example of a molecular marker assay) is reducing while the cost of phenotyping is increasing particularly in developed countries thus increasing the attractiveness of MAS as the development of the technology continues.

Steps for MAS

Generally the first step is to map the gene or quantitative trait locus (QTL) of interest first by using different techniques and then use this information for marker assisted selection. Generally, the markers to be used should be close to gene of interest (<5 recombination unit or cM) in order to ensure that only minor fraction of the selected individuals will be recombinants. Generally, not only a single marker but rather two markers are used in order to reduce the chances of an error due to homologous recombination. For example, if two flanking markers are used at same time with an interval between them of approximately 20cM, there is higher probability (99%) for recovery of the target gene.

QTL Mapping Techinques

In plants QTL mapping is generally achieved using bi-parental cross populations; a cross between two parents which have a contrasting phenotype for the trait of interest are developed. Commonly used populations are recombinant inbred lines (RILs), doubled haploids (DH), back cross and F_2. Linkage between the phenotype and markers which have already been mapped is tested in these populations in order to determine the position of the QTL. Such techniques are based on linkage and are therefore referred to as "linkage mapping".

Single Step MAS and QTL Mapping

In contrast to two-step QTL mapping and MAS, a single-step method for breeding typical plant populations has been developed. In such an approach, in the first few breeding cycles, markers linked to the trait of interest are identified by QTL mapping and later the same information in used in the same population. In this approach, pedigree structure are created from families that are created by crossing number of parents (in three-way or four way crosses).

Both phenotyping and genotyping is done using molecular markers mapped the possible location of QTL of interest. This will identify markers and their favorable alleles. Once these favorable marker alleles are identified, the frequency of such alleles will be increased and response to marker assisted selection is estimated. Marker allele(s) with desirable effect will be further used in next selection cycle or other experiments.

Paratransgenesis

Paratransgenesis is a technique that attempts to eliminate a pathogen from vector populations through transgenesis of a symbiont of the vector. The goal of this technique is to control vector-borne diseases.

The first step is to identify proteins that prevent the vector species from transmitting the pathogen. The genes coding for these proteins are then introduced into the symbiont, so that they can be expressed in the vector. The final step in the strategy is to introduce these transgenic symbionts into vector populations in the wild.

The first example of this technique used *Rhodnius prolixus* which is associated with the symbiont *Rhodococcus rhodnii. Rhodnius prolixus* is an important insect vector of Chagas's disease that is caused by *Trypanosoma cruzi*. The strategy was to engineer *R. rhodnii* to express proteins such as Cecropin A that are toxic to *Trypanosoma cruzi* or that block the transmission of *Trypanosoma cruzi*.

3

Applications of Recombinant DNA Technology

Recombinant DNA is widely used in biotechnology, medicine and research. Today, recombinant proteins and other products that result from the use of rDNA technology are found in essentially every western pharmacy, doctor's or veterinarian's office, medical testing laboratory, and biological research laboratory. In addition, organisms that have been manipulated using recombinant DNA technology, and products derived from those organisms have found their way into many farms, supermarkets, home medicine cabinets and even pet shops.

The most common application of recombinant DNA is in basic research, where it is important to most current work in the biological and biomedical sciences. Recombinant DNA is used to identify, map and sequence genes, and to determine their function. rDNA probes are employed in analyzing gene expression within individual cells, and throughout the tissues of whole organisms. Recombinant proteins are widely used as reagents in laboratory experiments and to generate antibody probes for examining protein synthesis within cells and organisms.

Many additional practical applications of recombinant DNA are found in human and veterinary medicine, in agriculture, and in bioengineering. Some specific examples are identified below.

- Recombinant human insulin. Recombinant insulin has almost completely replaced insulin obtained from animal sources (e.g. pigs and cattle) for the treatment of insulin-dependent diabetes. A variety of different recombinant insulin preparations are in widespread use. DrugBank entry.

- Recombinant human growth hormone (HGH, somatotropin). Growth hormone is administered to patients whose pituitary glands generate insufficient quantities to support normal growth and development. Before recombinant HGH became available, HGH for therapeutic use was obtained from pituitary glands of cadavers. This unsafe practice led to some patients developing Creutzfeldt-Jacob disease. Recombinant HGH eliminated this problem, and is now used therapeutically. It has also been misused as a performance enhancing drug by athletes and others. DrugBank entry
- Recombinant blood clotting factor VIII. Recombinant factor VIII is a blood-clotting protein that is administered to patients with forms of the bleeding disorder hemophilia, who are unable to produce factor VIII in quantities sufficient to support normal blood coagulation. Before the development of recombinant factor VIII, the protein was obtained by processing large quantities of human blood from multiple donors, which carried a very high risk of transmission of blood borne infectious diseases, for example HIV and hepatitis B. DrugBank entry
- Recombinant hepatitis B vaccine. Prevention of hepatitis B infection is controlled through the use of a recombinant hepatitis B vaccine, which contains a form of the hepatitis B virus surface antigen that is produced in yeast cells. The development of the recombinant subunit vaccine was an important and necessary development because hepatitis B virus, unlike other common viruses such as polio virus, cannot be grown in vitro. Vaccine information from Hepatitis B Foundation
- Diagnosis of infection with HIV. Each of the three widely-used methods for diagnosing HIV infection has been developed using recombinant DNA. The antibody test (ELISA or western blot) uses a recombinant HIV protein to test for the presence of antibodies that the body has produced in response to an HIV infection. The DNA test looks for the presence of HIV genetic material using reverse transcriptase polymerase chain reaction (RT-PCR). Development of the RT-PCR test was made possible by the molecular cloning and sequence analysis of HIV genomes. HIV testing page from US Centers for Disease Control (CDC)
- Golden rice is a recombinant variety of rice that has been engineered to express the enzymes responsible for â-carotene biosynthesis. This variety of rice holds substantial promise for reducing the incidence of

vitamin A deficiency in the world's population. Golden rice is not currently in use, pending the resolution of intellectual property, environmental and nutritional issues.

- Herbicide-resistant crops Commercial varieties of important agricultural crops (including soy, maize/corn, sorghum, canola, alfalfa and cotton) have been developed which incorporate a recombinant gene that results in resistance to the herbicide glyphosate (trade name Roundup), and simplifies weed control by glyphosate application. These crops are in common commercial use in several countries.
- Insect-resistant crops. Bacillus thuringeiensis is a bacterium that naturally produces a protein (Bt toxin) with insecticidal properties. The bacterium has been applied to crops as an insect-control strategy for many years, and this practice has been widely adopted in agriculture and gardening. Recently, plants have been developed which express a recombinant form of the bacterial protein, which may effectively control some insect predators. Environmental issues associated with the use of these transgenic crops have not been fully resolved.

History of Recombinant DNA

The idea for recombinant DNA was first proposed by Peter Lobban, a graduate student of Prof. Dale Kaiser in the Biochemistry Department at Stanford University Medical School. The first publications describing the successful production and intracellular replication of recombinant DNA appeared in 1972 and 1973. Stanford University applied for a US patent on recombinant DNA in 1974, listing the inventors as Stanley N. Cohen and Herbert W. Boyer; this patent was awarded in 1980. The first licensed drug generated using recombinant DNA technology was human insulin, developed by Genentech and Licensed by Eli Lilly and Company.

Controversy

Scientists associated with the initial development of recombinant DNA methods recognized that the potential existed for organisms containing recombinant DNA to have undesirable or dangerous properties. At the 1975 Asilomar Conference on Recombinant DNA, these concerns were discussed and a voluntary moratorium on recombinant DNA research was initiated for experiments that were thought to be particularly risky. This moratorium was widely observed until the National Institutes of Health (USA) developed and issued formal guidelines for rDNA work. Today, recombinant DNA molecules and recombinant proteins are usually not regarded as dangerous. However, concerns remain about some organisms that express recombinant DNA, particularly when they leave the laboratory and are introduced into the environment or food chain. These concerns are discussed in the articles on genetically-modified organisms and genetically-modified food controversies.

Formal Genetics

Genetics a discipline of biology, is the science of genes, heredity, and variation in living organisms. Genetics deals with the molecular structure and function of genes, with gene behavior in the context of a cell or organism (e.g. dominance and epigenetics), with patterns of inheritance from parent to offspring, and with gene distribution, variation and change in populations. Given that genes are universal to living organisms, genetics can be applied to the study of all living systems, from viruses and bacteria, through plants (especially crops) and domestic animals, to humans (as in medical genetics).

The fact that living things inherit traits from their parents has been used since prehistoric times to improve crop plants and animals through selective breeding. However, the modern science of genetics, which seeks to understand the process of inheritance, only began with the work of Gregor Mendel in the mid-19th century. Although he did not know the physical basis for heredity, Mendel observed that organisms inherit traits via discrete units of inheritance, which are now called genes.

Genes correspond to regions within DNA, a molecule composed of a chain of four different types of nucleotides—the sequence of these nucleotides is the genetic information organisms inherit. DNA naturally occurs in a double stranded form, with nucleotides on each strand complementary to each other. Each strand can act as a template for creating a new partner strand. This is the physical method for making copies of genes that can be inherited.

The sequence of nucleotides in a gene is translated by cells to produce a chain of amino acids, creating proteins—the order of amino acids in a protein corresponds to the order of nucleotides in the gene. This relationship between nucleotide sequence and amino acid sequence is known as the genetic code. The amino acids in a protein determine how it folds into a three-dimensional shape; this structure is, in turn, responsible for the protein's function. Proteins carry out almost all the functions needed for cells to live. A change to the DNA in a gene can change a protein's amino acids, changing its shape and function: this can have a dramatic effect in the cell and on the organism as a whole.

Although genetics plays a large role in the appearance and behavior of organisms, it is the combination of genetics with what an organism experiences that determines the ultimate outcome. For example, while genes play a role in determining an organism's size, the nutrition and health it experiences after inception also have a large effect.

History

DNA, the molecular basis for inheritance. Each strand of DNA is a chain of nucleotides, matching each other in the center to form what look like rungs on a twisted ladder.

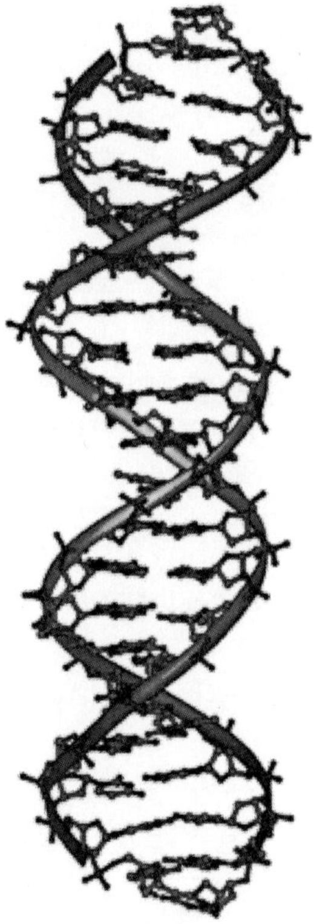

Although the science of genetics began with the applied and theoretical work of Gregor Mendel in the mid-19th century, other theories of inheritance preceded Mendel. A popular theory during Mendel's time was the concept of blending inheritance: the idea that individuals inherit a smooth blend of traits from their parents. Mendel's work disproved this, showing that traits are composed of combinations of distinct genes rather than a continuous blend. Another theory that had some support at that time was the inheritance of acquired characteristics: the belief that individuals inherit traits strengthened by their parents.

This theory (commonly associated with Jean-Baptiste Lamarck) is now known to be wrong—the experiences of individuals do not affect the genes they pass to their children. Other theories included the pangenesis of Charles Darwin (which had both acquired and inherited aspects) and Francis Galton's reformulation of pangenesis as both particulate and inherited.

Mendelian and Classical Genetics

Modern genetics started with Gregor Johann Mendel, a German-Czech Augustinian monk and scientist who studied the nature of inheritance in plants. In his paper "Versuche über Pflanzenhybriden" ("Experiments on Plant Hybridization"), presented in 1865 to the Naturforschender Verein (Society for Research in Nature) in Brünn, Mendel traced the inheritance patterns of certain traits in pea plants and described them mathematically. Although this pattern of inheritance could only be observed for a few traits, Mendel's work suggested that heredity was particulate, not acquired, and that the inheritance patterns of many traits could be explained through simple rules and ratios.

The importance of Mendel's work did not gain wide understanding until the 1890s, after his death, when other scientists working on similar problems re-discovered his research. William Bateson, a proponent of Mendel's work, coined the word genetics in 1905. (The adjective genetic, derived from the Greek word genesis—ãÝíåóéò, "origin", predates the noun and was first used in a biological sense in 1860.) Bateson popularized the usage of the word genetics to describe the study of inheritance in his inaugural address to the Third International Conference on Plant Hybridization in London, England, in 1906. After the rediscovery of Mendel's work, scientists tried to determine which molecules in the cell were responsible for inheritance. In 1910, Thomas Hunt Morgan argued that genes are on chromosomes, based on observations of a sex-linked white eye mutation in fruit flies.

In 1913, his student Alfred Sturtevant used the phenomenon of genetic linkage to show that genes are arranged linearly on the chromosome.

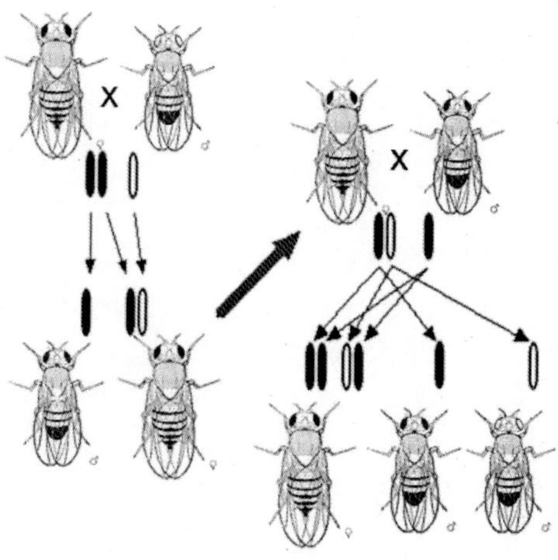

Figure: *Morgan's observation of sex-linked inheritance of a mutation causing white eyes in Drosophila led him to the hypothesis that genes are located upon chromosomes.*

Molecular Genetics

Although genes were known to exist on chromosomes, chromosomes are composed of both protein and DNA—scientists did not know which of these is responsible for inheritance. In 1928, Frederick Griffith discovered the phenomenon of transformation (Griffith's experiment): dead bacteria could transfer genetic material to "transform" other still-living bacteria. Sixteen years later, in 1944, Oswald Theodore Avery, Colin McLeod and Maclyn McCarty identified the molecule responsible for transformation as DNA. The Hershey-Chase experiment in 1952 also showed that DNA (rather than protein) is the genetic material of the viruses that infect bacteria, providing further evidence that DNA is the molecule responsible for inheritance.

James D. Watson and Francis Crick determined the structure of DNA in 1953, using the X-ray crystallography work of Rosalind Franklin and Maurice Wilkins that indicated DNA had a helical structure (i.e., shaped like a corkscrew).

Their double-helix model had two strands of DNA with the nucleotides pointing inward, each matching a complementary nucleotide on the other strand to form what looks like rungs on a twisted ladder. This structure showed that genetic information exists in the sequence of nucleotides on each strand of DNA. The structure also suggested a simple method for duplication: if the strands are separated, new partner strands can be reconstructed for each based on the sequence of the old strand.

Although the structure of DNA showed how inheritance works, it was still not known how DNA influences the behavior of cells. In the following years, scientists tried to understand how DNA controls the process of protein production. It was discovered that the cell uses DNA as a template to create matching messenger RNA (a molecule with nucleotides, very similar to DNA). The nucleotide sequence of a messenger RNA is used to create an amino acid sequence in protein; this translation between nucleotide and amino acid sequences is known as the genetic code.

With this molecular understanding of inheritance, an explosion of research became possible. One important development was chain-termination DNA sequencing in 1977 by Frederick Sanger. This technology allows scientists to read the nucleotide sequence of a DNA molecule. In 1983, Kary Banks Mullis developed the polymerase chain reaction, providing a quick way to isolate and amplify a specific section of a DNA from a mixture. Through the pooled efforts of the Human Genome Project and the parallel private effort by Celera Genomics, these and other methods culminated in the sequencing of the human genome in 2003.

Features of Inheritance

Discrete Inheritance and Mendel's Laws

At its most fundamental level, inheritance in organisms occurs by means of discrete traits, called genes. This property was first observed by Gregor Mendel, who studied the segregation of heritable traits in pea plants. In his experiments studying the trait for flower color, Mendel observed that the flowers of each pea plant were either purple or white—but never an intermediate between the two colors. These different, discrete versions of the same gene are called alleles.

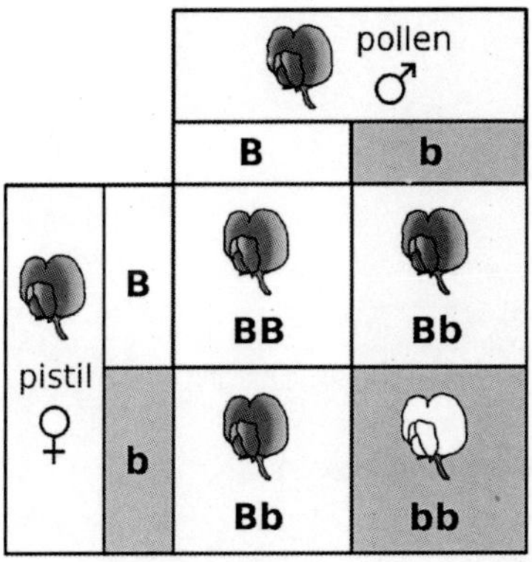

Figure: *A Punnett square depicting a cross between two pea plants heterozygous for purple (B) and white (b) blossoms*

In the case of pea, which is a diploid species, each individual plant has two copies of each gene, one copy inherited from each parent. Many species, including humans, have this pattern of inheritance. Diploid organisms with two copies of the same allele of a given gene are called homozygous at that gene locus, while organisms with two different alleles of a given gene are called heterozygous.

The set of alleles for a given organism is called its genotype, while the observable traits of the organism are called its phenotype. When organisms are heterozygous at a gene, often one allele is called dominant as its qualities dominate the phenotype of the organism, while the other allele is called recessive as its qualities recede and are not observed. Some alleles do not have complete dominance and instead have incomplete dominance by expressing an intermediate phenotype, or codominance by expressing both alleles at once. When a pair of organisms reproduce sexually, their offspring randomly inherit one of the two alleles from each parent. These observations of discrete inheritance and the segregation of alleles are collectively known as Mendel's first law or the Law of Segregation.

Notation and Diagrams

Geneticists use diagrams and symbols to describe inheritance. A gene is represented by one or a few letters. Often a "+" symbol is used to mark the usual, non-mutant allele for a gene.

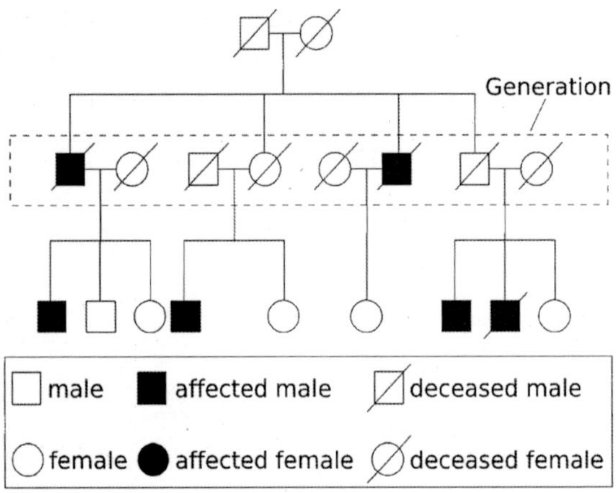

Figure: *Genetic pedigree charts help track the inheritance patterns of traits.*

In fertilization and breeding experiments (and especially when discussing Mendel's laws) the parents are referred to as the "P" generation and the offspring as the "F1" (first filial) generation. When the F1 offspring mate with each other, the offspring are called the "F2" (second filial) generation. One of the common diagrams used to predict the result of cross-breeding is the Punnett square. When studying human genetic diseases, geneticists often use pedigree charts to represent the inheritance of traits. These charts map the inheritance of a trait in a family tree.

Interactions of Multiple Genes

Human height is a trait with complex genetic causes. Francis Galton's data from 1889 shows the relationship between offspring height as a function of mean parent height. While correlated, remaining variation in offspring heights indicates environment is also an important factor in this trait. Organisms have thousands of genes, and in sexually reproducing organisms these genes generally assort independently of each other.

This means that the inheritance of an allele for yellow or green pea color is unrelated to the inheritance of alleles for white or purple flowers. This phenomenon, known as "Mendel's second law" or the "Law of independent assortment", means that the alleles of different genes get shuffled between parents to form offspring with many different combinations. Often different genes can interact in a way that influences the same trait. In the Blue-eyed

Mary (Omphalodes verna), for example, there exists a gene with alleles that determine the color of flowers: blue or magenta. Another gene, however, controls whether the flowers have color at all or are white. When a plant has two copies of this white allele, its flowers are white—regardless of whether the first gene has blue or magenta alleles. This interaction between genes is called epistasis, with the second gene epistatic to the first.

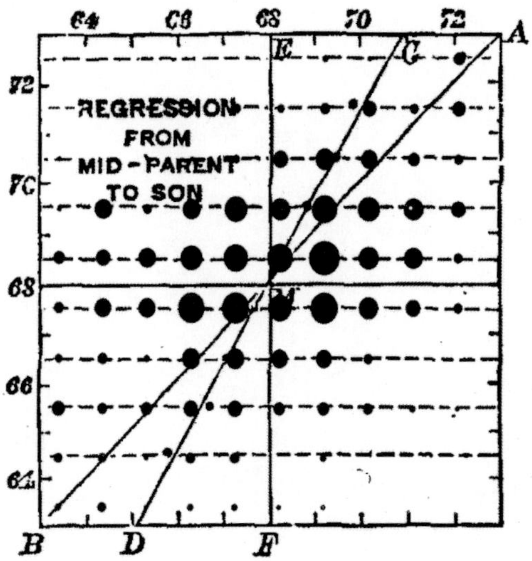

Many traits are not discrete features (e.g. purple or white flowers) but are instead continuous features (e.g. human height and skin color). These complex traits are products of many genes. The influence of these genes is mediated, to varying degrees, by the environment an organism has experienced. The degree to which an organism's genes contribute to a complex trait is called heritability. Measurement of the heritability of a trait is relative—in a more variable environment, the environment has a bigger influence on the total variation of the trait. For example, human height is a trait with complex causes. It has a heritability of 89% in the United States. In Nigeria, however, where people experience a more variable access to good nutrition and health care, height has a heritability of only 62%.

Molecular Basis for Inheritance

DNA and Chromosomes

The molecular structure of DNA. Bases pair through the arrangement of hydrogen bonding between the strands.

The molecular basis for genes is deoxyribonucleic acid (DNA). DNA is composed of a chain of nucleotides, of which there are four types: adenine

(A), cytosine (C), guanine (G), and thymine (T). Genetic information exists in the sequence of these nucleotides, and genes exist as stretches of sequence along the DNA chain. Viruses are the only exception to this rule—sometimes viruses use the very similar molecule RNA instead of DNA as their genetic material.

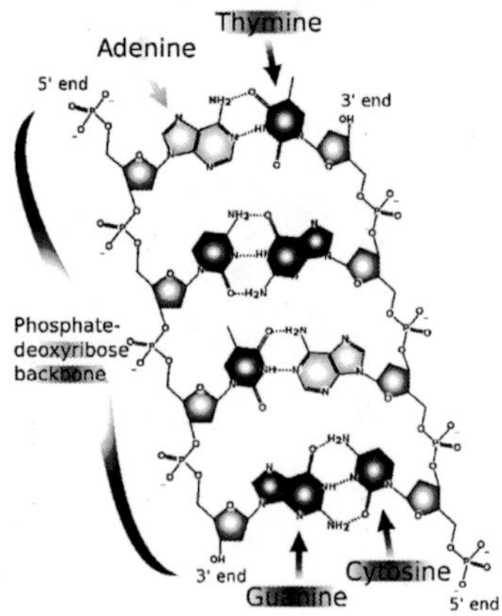

DNA normally exists as a double-stranded molecule, coiled into the shape of a double-helix. Each nucleotide in DNA preferentially pairs with its partner nucleotide on the opposite strand: A pairs with T, and C pairs with G.

Thus, in its two-stranded form, each strand effectively contains all necessary information, redundant with its partner strand. This structure of DNA is the physical basis for inheritance:

DNA replication duplicates the genetic information by splitting the strands and using each strand as a template for synthesis of a new partner strand. Genes are arranged linearly along long chains of DNA base-pair sequences.

In bacteria, each cell usually contains a single circular genophore, while eukaryotic organisms (including plants and animals) have their DNA arranged in multiple linear chromosomes.

These DNA strands are often extremely long; the largest human chromosome, for example, is about 247 million base pairs in length. The DNA of a chromosome is associated with structural proteins that organize, compact, and control access to the DNA, forming a material called chromatin; in eukaryotes, chromatin is usually composed of nucleosomes, segments of DNA wound around cores of histone proteins. The full set of hereditary material in an organism (usually the combined DNA sequences of all chromosomes) is called the genome.

While haploid organisms have only one copy of each chromosome, most animals and many plants are diploid, containing two of each chromosome and thus two copies of every gene.

The two alleles for a gene are located on identical loci of sister chromatids, each allele inherited from a different parent.

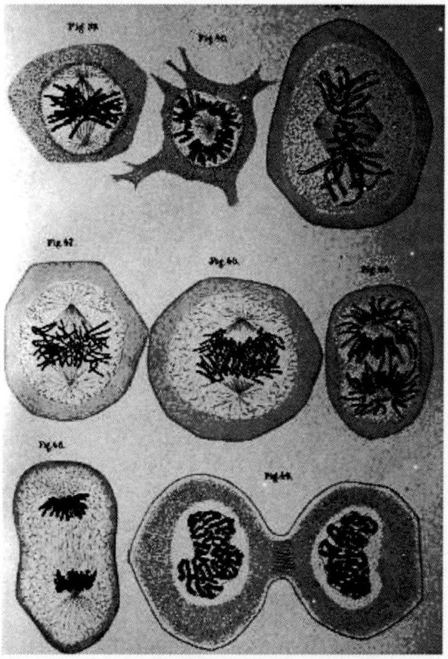

Walther Flemming's 1882 diagram of eukaryotic cell division. Chromosomes are copied, condensed, and organized. Then, as the cell divides, chromosome copies separate into the daughter cells.

Many species have so called sex chromosomes. They are special in that they determine the sex of the organism. In humans and many other animals, the Y-chromosome contains the gene that triggers the development of the specifically male characteristics. In evolution, this chromosome has lost most of its content and also most of its genes, while the X chromosome is similar to the other chromosomes and contains many genes. The X and Y chromosomes form a very heterogeneous pair before cell division.

Reproduction

When cells divide, their full genome is copied and each daughter cell inherits one copy. This process, called mitosis, is the simplest form of reproduction and is the basis for asexual reproduction. Asexual reproduction can also occur in multicellular organisms, producing offspring that inherit their genome from a single parent. Offspring that are genetically identical to their parents are called clones.

Eukaryotic organisms often use sexual reproduction to generate offspring that contain a mixture of genetic material inherited from two different parents. The process of sexual reproduction alternates between forms that contain single copies of the genome (haploid) and double copies (diploid). Haploid cells fuse and combine genetic material to create a diploid cell with paired chromosomes. Diploid organisms form haploids by dividing, without replicating their DNA, to create daughter cells that randomly inherit one of each pair of chromosomes. Most animals and many plants are diploid for most of their lifespan, with the haploid form reduced to single cell gametes such as sperm or eggs.

Although they do not use the haploid/diploid method of sexual reproduction, bacteria have many methods of acquiring new genetic information. Some bacteria can undergo conjugation, transferring a small circular piece of DNA to another bacterium. Bacteria can also take up raw DNA fragments found in the environment and integrate them into their genomes, a phenomenon known as transformation. These processes result in horizontal gene transfer, transmitting fragments of genetic information between organisms that would be otherwise unrelated.

Recombination and Linkage

Thomas Hunt Morgan's 1916 illustration of a double crossover between chromosomes the diploid nature of chromosomes allows for genes on different chromosomes to assort independently during sexual reproduction, recombining to form new combinations of genes. Genes on the same chromosome would theoretically never recombine, however, were it not for the process of chromosomal crossover. During crossover, chromosomes exchange stretches of DNA, effectively shuffling the gene alleles between the chromosomes. This process of chromosomal crossover generally occurs during meiosis, a series of cell divisions that creates haploid cells.

The probability of chromosomal crossover occurring between two given points on the chromosome is related to the distance between the points. For an arbitrarily long distance, the probability of crossover is high enough that the inheritance of the genes is effectively uncorrelated. For genes that are closer together, however, the lower probability of crossover means that the genes demonstrate genetic linkage—alleles for the two genes tend to be inherited together. The amounts of linkage between a series of genes can be combined to form a linear linkage map that roughly describes the arrangement of the genes along the chromosome.

4

Testing of DNA and RNA

The nucleic acids are the building blocks of living organisms. You may have heard of DNA described the same way. Guess what? DNA is just one type of nucleic acid. Some other types are RNA, mRNA, and tRNA. All of these "Na's" work together to help cells replicate and build proteins.

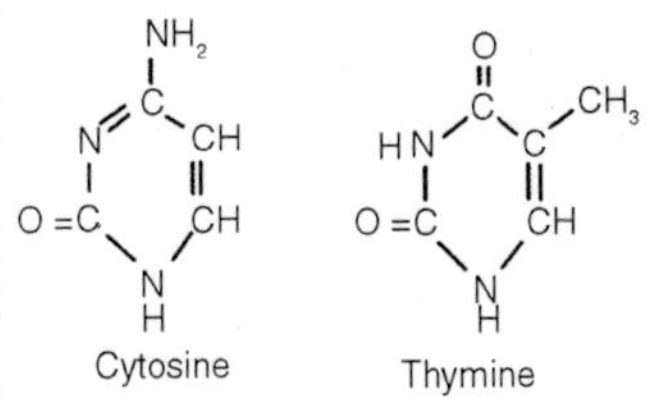

Pyrimidine Bases

Basics

We already told you about the biggie nucleic acids (DNA, RNA, mRNA, tRNA). They are actually made up chains of base pairs stretching from only a few to millions. When those pairs combine in super long chains (DNA), they make a shape called a double helix. The double helix shape is like a twisty ladder. The base pairs are the rungs. We're very close to talking about the biology of cells here. Back to the chemistry..

Five Easy Pieces

There are five easy parts of nucleic acids. All nucleic acids are made up of the same building blocks (monomers). Chemists call the monomers nucleotides. The five pieces are Uracil, Cytosine, Thymine, Adenine, and Guanine. Just as there are twenty (20) amino acids needed by humans to survive, there are five (5)nucleotides.

These nucleotides are made of three parts:

1. A five carbon sugar
2. A base that has a nitrogen (N) atom
3. An ion of phosphoric acid

The first isolation of what we now refer to as DNA was accomplished by Johann Friedrich Miescher *circa* 1870. He reported finding a weakly acidic substance of unknown function in the nuclei of human white blood cells, and named this material "nuclein". A few years later, Miescher separated nuclein into protein and nucleic acid components. In the 1920's nucleic acids were found to be major components of chromosomes, small gene-carrying bodies in the nuclei of complex cells. Elemental analysis of nucleic acids showed the presence of phosphorus, in addition to the usual C, H, N & O. Unlike proteins, nucleic acids contained no sulphur. Complete hydrolysis of chromosomal nucleic acids gave inorganic phosphate, 2-deoxyribose (a previously unknown sugar) and four different heterocyclic bases (shown in the following diagram). To reflect the unusual sugar component, chromosomal nucleic acids are called deoxyribonucleic acids, abbreviated DNA. Analogous nucleic acids in which the sugar component is ribose are termed ribonucleic acids, abbreviated RNA. The acidic character of the nucleic acids was attributed to the phosphoric acid moiety.

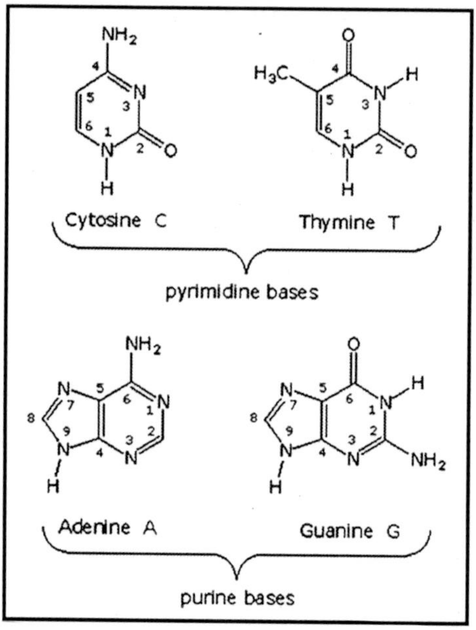

The two monocyclic bases shown here are classified as pyrimidines, and the two bicyclic bases are purines. Each has at least one N-H site at which an organic substituent may be attached. They are all polyfunctional bases, and may exist in tautomeric forms.

Base-catalyzed hydrolysis of DNA gave four nucleoside products, which proved to be N-glycosides of 2'-deoxyribose combined with the heterocyclic amines. Structures and names for these nucleosides will be displayed above by clicking on the heterocyclic base diagram. The base components are coloured green, and the sugar is black. As noted in the 2'-deoxycytidine structure on the left, the numbering of the sugar carbons makes use of primed numbers to distinguish them from the heterocyclic base sites. The corresponding N-glycosides of the common sugar ribose are the building blocks of RNA, and are named adenosine, cytidine, guanosine and uridine (a thymidine analog missing the methyl group).

From this evidence, nucleic acids may be formulated as alternating copolymers of phosphoric acid (P) and nucleosides (N), as shown:

$$\sim P - N - P - N' - P - N'' - P - N''' - P - N \sim$$

At first the four nucleosides, distinguished by prime marks in this crude formula, were assumed to be present in equal amounts, resulting in a uniform structure, such as that of starch. However, a compound of this kind, presumably common to all organisms, was considered too simple to hold the hereditary information known to reside in the chromosomes. This view was challenged in 1944, when Oswald Avery and colleagues demonstrated that bacterial DNA was likely the geneic agent that carried information from one organism to another in a process called "transformation".

He concluded that*"nucleic acids must be regarded as possessing biological specificity, the chemical basis of which is as yet undetermined."* Despite this finding, many scientists continued to believe that chromosomal proteins, which differ across species, between individuals, and even within a given organism, were the locus of an organism's genetic information.

It should be noted that single celled organisms like bacteria do not have a well-defined nucleus. Instead, their single chromosome is associated with specific roteins in a region called a "nucleoid". Nevertheless, the DNA from bacteria has the same composition and general structure as that from multicellular organisms, including human beings.

Views about the role of DNA in inheritance changed in the late 1940's and early 1950's. By conducting a careful analysis of DNA from many sources, Erwin Chargaff found its composition to be species specific. In addition, he found that the amount of adenine (A) always equaled the amount of thymine

(T), and the amount of guanine (G) always equaled the amount of cytosine (C), regardless of the DNA source. As set forth in the following table, the ratio of (A+T) to (C+G) varied from 2.70 to 0.35. The last two organisms are bacteria.

Nucleoside Base Distribution in DNA

In a second critical study, Alfred Hershey and Martha Chase showed that when a bacterium is infected and genetically transformed by a virus, at least 80% of the viral DNA enters the bacterial cell and at least 80% of the viral protein remains outside. Together with the Chargaff findings this work established DNA as the repository of the unique genetic characteristics of an organism.

Organism	Base Composition (mole %)				Base Ratios		Ratio (A+T)/(G+C)
	A	G	T	C	A/T	G/C	
Human	30.9	19.9	29.4	19.8	1.05	1.00	1.52
Chicken	28.8	20.5	29.2	21.5	1.02	0.95	1.38
Yeast	31.3	18.7	32.9	17.1	0.95	1.09	1.79
Clostridium perfringens	36.9	14.0	36.3	12.8	1.01	1.09	2.70
Sarcina lutea	13.4	37.1	12.4	37.1	1.08	1.00	0.35

The Chemical Nature of DNA

The polymeric structure of DNA may be described in terms of monomeric units of increasing complexity. In the top shaded box of the following illustration, the three relatively simple components mentioned earlier are shown. Below that on the left, formulas for phosphoric acid and a nucleoside are drawn.Condensation polymerization of these leads to the DNA formulation outlined above.

Finally, a 5'- monophosphate ester, called a nucleo-tide may be drawn as a single monomer unit, shown in the shaded box to the right. Since a monophosphate ester of this kind is a strong acid (pK_a of 1.0), it will be fully ionized at the usual physiological pH (ca.7.4).

Names for these DNA components are given in the table to the right of the diagram. Isomeric 3'-monophospate nucleotides are also known, and both isomers are found in cells.

They may be obtained by selective hydrolysis of DNA through the action of nuclease enzymes. Anhydride-like di- and tri-phosphate nucleotides have been identified as important energy carriers in biochemical reactions, the most common being ATP (adenosine 5'-triphosphate).

Names of DNA Base Derivatives

Base	Nucleoside	5'-Nucleotide
Adenine	2'-Deoxyadenosine	2'-Deoxyadenosine-5'-monophosphate
Cytosine	2'-Deoxycytidine	2'-Deoxycytidine-5'-monophosphate
Guanine	2'-Deoxyguanosine	2'-Deoxyguanosine-5'-monophosphate
Thymine	2'-Deoxythymidine	2'-Deoxythymidine-5'-monophosphate

A complete structural representation of a segment of the DNA polymer formed from 5'-nucleotides. Several important characteristics of this formula should be noted.

- First, the remaining P-OH function is quite acidic and is completely ionized in biological systems.
- Second, the polymer chain is structurally directed. One end is different from the other.
- Third, although this appears to be a relatively simple polymer, the possible permutations of the four nucleosides in the chain become very large as the chain lengthens.
- Fourth, the DNA polymer is much larger than originally believed. Molecular weights for the DNA from multicellular organisms are commonly 10^9 or greater.

Information is stored or encoded in the DNA polymer by the pattern in which the four nucleotides are arranged. To access this information the pattern must be "read" in a linear fashion, just as a bar code is read at a supermarket checkout. Because living organisms are extremely complex, a correspondingly large amount of information related to this complexity must be stored in the DNA.

Consequently, the DNA itself must be very large, as noted above. Even the single DNA molecule from an *E. coli* bacterium is found to have roughly a million nucleotide units in a polymer strand, and would reach a millimeter in length if stretched out. The nuclei of multicellular organisms incorporate chromosomes, which are composed of DNA combined with nuclear proteins called histones.

The fruit fly has 8 chromosomes, humans have 46 and dogs 78 (note that the amount of DNA in a cell's nucleus does not correlate with the number of chromosomes). The DNA from the smallest human chromosome is over ten times larger than *E. coli* DNA, and it has been estimated that the total DNA in a human cell would extend to 2 meters in length if unraveled. Since the nucleus is only about 5ìm in diameter, the chromosomal DNA must be packed tightly to fit in that small volume.

In addition to its role as a stable informational library, chromosomal DNA must be structured or organized in such a way that the chemical machinery of the cell will have easy access to that information, in order to make important molecules such as polypeptides. Furthermore, accurate copies of the DNA code must be created as cells divide, with the replicated DNA molecules passed on to subsequent cell generations, as well as to progeny of the organism.

RNA, a Different Nucleic Acid

The high molecular weight nucleic acid, DNA, is found chiefly in the nuclei of complex cells, known as eucaryotic cells, or in the nucleoid regions ofprocaryotic cells, such as bacteria. It is often associated with proteins that help to pack it in a usable fashion. In contrast, a lower molecular weight, but much more abundant nucleic acid, RNA, is distributed throughout the cell, most commonly in small numerous organelles called ribosomes.

Three kinds of RNA are identified, the largest subgroup (85 to 90%) being ribosomal RNA, rRNA, the major component of ribosomes, together with proteins. The size of rRNA molecules varies, but is generally less than a thousandth the size of DNA. The other forms of RNA are messenger RNA, mRNA, and transfer RNA, tRNA. Both have a more transient existence and are smaller than rRNA.

All these RNA's have similar constitutions, and differ from DNA in two important respects. As shown in the following diagram, the sugar component of RNA is ribose, and the pyrimidine base uracil replaces the thymine base of DNA.

The RNA's play a vital role in the transfer of information (transcription) from the DNA library to the protein factories called ribosomes, and in the interpretation of that information (translation) for the synthesis of specific polypeptides. These functions will be described later. A complete structural representation of a segment of the RNA polymer formed from 5'-nucleotides.

The Secondary Structure of DNA

In the early 1950's the primary structure of DNA was well established, but a firm understanding of its secondary structure was lacking. Indeed, the situation was similar to that occupied by the proteins a decade earlier, before the alpha helix and pleated sheet structures were proposed by Linus Pauling.

Many researchers grappled with this problem, and it was generally conceded that the molar equivalences of base pairs (A & T and C & G) discovered by Chargaff would be an important factor. A scientist, working at King's College, obtained X-ray diffraction evidence that suggested a long helical structure of uniform thickness.

The Double Helix

After many trials and modifications, an ingenious double helix model for the secondary structure of DNA. Two strands of DNA were aligned anti-parallel to each other, *i.e.* with opposite 3' and 5' ends, as shown in part a of the following diagram. Complementary primary nucleotide structures for each strand allowed intra-strand hydrogen bonding between each pair of bases. These complementary strands are coloured red and green in the diagram. Coiling these coupled strands then leads to a double helix structure, shown as cross-linked ribbons in part b of the diagram. A space-filling molecular model of a short segment is displayed in part c on the right.

The helix shown here has ten base pairs per turn, and rises 3.4 Å in each turn. This right-handed helix is the favoured conformation in aqueous systems, and has been termed the B-helix. Two alternating grooves are evident, a wide and deep major groove, and a shallow and narrow minor groove. Other molecules, including polypeptides, may insert into these grooves, and in so doing perturb the chemistry of DNA. Other helical structures of DNA have also been observed, and are designated by letters (*e.g.* A and Z).

The Double Helix Structure for DNA

DNA Replication: In their announcement of a double helix structure for DNA, It is stated, *"It has not escaped our notice that the specific pairing we have postulated immediately suggests a possible copying mechanism for the genetic material."*

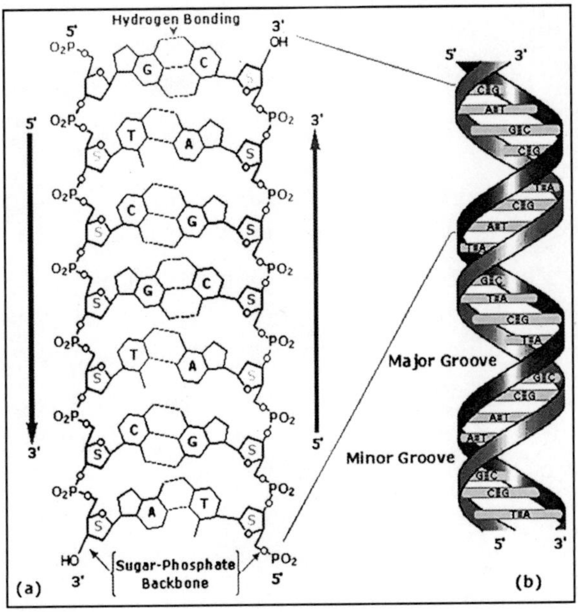

The essence of this suggestion is that, if separated, each strand of the molecule might act as a template on which a new complementary strand might be assembled, leading finally to two identical DNA molecules. Indeed, replication does take place in this fashion when cells divide, but the events leading up to the actual synthesis of complementary DNA strands are sufficiently complex that they will not be described in any detail.

As depicted in the following drawing, the DNA of a cell is tightly packed into chromosomes. First, the DNA is wrapped around small proteins called histones. These bead-like structures are then further organized and folded into chromatin aggregates that make up the chromosomes.

An overall packing efficiency of 7,000 or more is thus achieved. Clearly a sequence of unfolding events must take place before the information encoded in the DNA can be used or replicated.

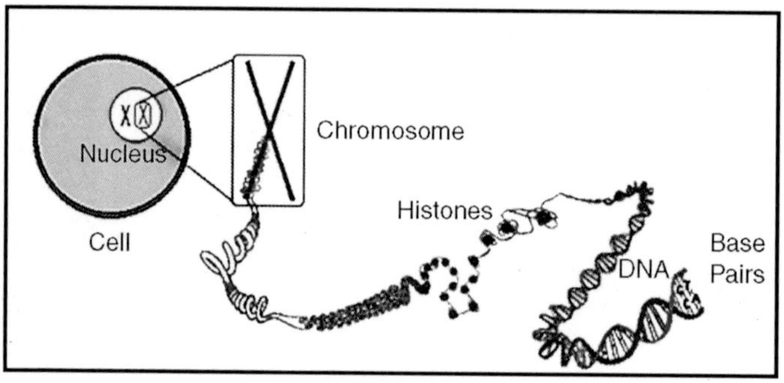

Once the double stranded DNA is exposed, a group of enzymes act to accomplish its replication.

These are described briefly here:

- *Topoisomerase:* This enzyme initiates unwinding of the double helix by cutting one of the strands.
- *Helicase:* This enzyme assists the unwinding. Note that many hydrogen bonds must be broken if the strands are to be separated.
- *SSB:* A single-strand binding-protein stabilizes the separated strands, and prevents them from recombining, so that the polymerization chemistry can function on the individual strands.
- *DNA Polymerase:* This family of enzymes link together nucleotide triphosphate monomers as they hydrogen bond to complementary bases. These enzymes also check for errors (roughly ten per billion), and make corrections.
- *Ligase:* Small unattached DNA segments on a strand are united by this enzyme.

Polymerization of nucleotides takes place by the phosphorylation reaction described by the following equation.

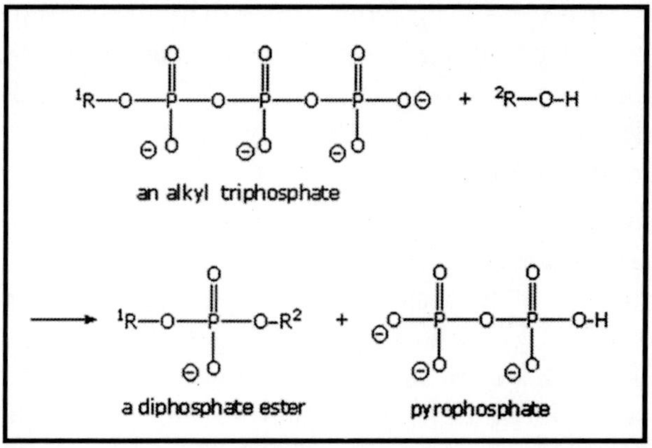

Di- and triphosphate esters have anhydride-like structures and are consequently reactive phosphorylating reagents, just as carboxylic anhydrides are acylating reagents. Since the pyrophosphate anion is a better leaving group than phosphate, triphosphates are more powerful phosphorylating agents than are diphosphates.

Formulas for the corresponding 5'-derivatives of adenosine will be displayed, and similar derivatives exist for the other three common nucleosides. The DNA polymerization process that builds the complementary strands in replication, could in principle take place in two ways. Referring to the general equation above, R^1 could represent the next nucleotide unit to be attached to the growing DNA strand, with R^2 being this strand. Alternatively, these assignments could be reversed.

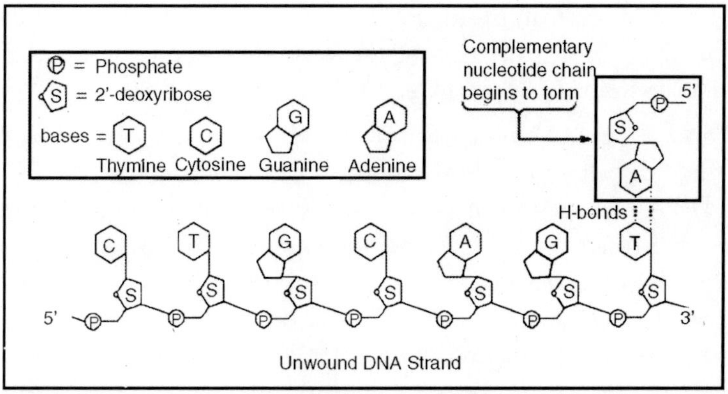

In practice, the former proves to be the best arrangement. Since triphosphates are very reactive, the lifetime of such derivatives in an aqueous environment is relatively short. However, such derivatives of the individual nucleosides are repeatedly synthesized by the cell for a variety of purposes, providing a steady supply of these reagents. In contrast, the growing DNA segment must maintain its functionality over the entire replication process, and can not afford to be changed by a spontaneous hydrolysis event. As a result, these chemical properties are best accomodated by a polymerization process that proceeds at the 3'-end of the growing strand by 5'-phosphorylation involving a nucleotide triphosphate.

The polymerization mechanism described here is constant. It always extends the developing DNA segment towards the 3'-end (*i.e.* when a nucleotide triphosphate attaches to the free 3'-hydroxyl group of the strand, a new 3'-hydroxyl is generated). There is sometimes confusion on this point, because the original DNA strand that serves as a template is read from the 3'-end towards the 5'-end, and authors may not be completely clear as to which terminology is used.

Because of the directional demand of the polymerization, one of the DNA strands is easily replicated in a continuous fashion, whereas the other strand can only be replicated in short segmental pieces. This is illustrated in the following diagram. Separation of a portion of the double helix takes place at a site called the replication fork. As replication of the separate strands occurs, the replication fork moves away (to the left in the diagram), unwinding additional lengths of DNA. Since the fork in the diagram is moving towards the 5'-end of the red-coloured strand, replication of this strand may take place in a continuous fashion (building the new green strand in a 5' to 3' direction). This continuously formed new strand is called the leading strand. In contrast, the replication fork moves towards the 3'-end of the original green strand, preventing continuous polymerization of a complementary new red strand. Short segments of complementary DNA, called Okazaki fragments, are produced, and these are linked together later by the enzyme ligase. This new DNA strand is called the lagging strand.

When you consider that a human cell has roughly 10^9 base pairs in its DNA, and may divide into identical daughter cells in 14 to 24 hours, the efficiency of DNA replication must be extraordinary. The procedure described above will replicate about 50 nucleotides per second, so there must be many thousand such replication sites in action during cell division. A given length of double stranded DNA may undergo strand unwinding at numerous sites in response to promoter actions. The unraveled "bubble" of single stranded DNA has two replication forks, so assembly of new complementary strands may proceed in two directions. The polymerizations associated with several such bubbles fuse together to achieve full replication of the entire DNA double helix. Note that the events shown proceed from top to bottom in the diagram.

Repair of DNA Damage and Replication Errors

One of the benefits of the double stranded DNA structure is that it lends itself to repair, when structural damage or replication errors occur.

Several kinds of chemical change may cause damage to DNA:

- Spontaneous hydrolysis of a nucleoside removes the heterocyclic base component.
- Spontaneous hydrolysis of cytosine changes it to a uracil.
- Various toxic metabolites may oxidize or methylate heterocyclic base components.
- Ultraviolet light may dimerize adjacent cytosine or thymine bases.

All these transformations disrupt base pairing at the site of the change, and this produces a structural deformation in the double helix.. Inspection-repair enzymes detect such deformations, and use the undamaged nucleotide at that site as a template for replacing the damaged unit. These repairs reduce errors in DNA structure from about one in ten million to one per trillion.

RNA and Protein Synthesis

The genetic information stored in DNA molecules is used as a blueprint for making proteins. Why proteins? Because these macromolecules have diverse primary, secondary and tertiary structures that equip them to carry out the numerous functions necessary to maintain a living organism.

As noted in the protein, these functions include:

- Structural integrity (hair, horn, eye lenses etc.).
- Molecular recognition and signaling (antibodies and hormones).
- Catalysis of reactions (enzymes)..
- Molecular transport (hemoglobin transports oxygen).
- Movement (pumps and motors).

The critical importance of proteins in life processes is demonstrated by numerous genetic diseases, in which small modifications in primary structure produce debilitating and often disasterous consequences.

Such genetic diseases incluse Tay-Sachs, phenylketonuria (PKU), sickel cell anemia, achondroplasia, and Parkinson disease. The unavoidable conclusion is that proteins are of central importance in living cells, and that proteins must therefore be continuously prepared with high structural fidelity by appropriate cellular chemistry. Early geneticists identifed genes as hereditary units that determined the appearance and/ or function of an organism (*i.e.* its phenotype). We now define genes as sequences of DNA that occupy specific locations on a chromosome. The original proposal that each gene controlled the formation of a single enzyme has since been modified as: one gene = one polypeptide. The intriguing question of how the information encoded in DNA is converted to the actual construction of a specific polypeptide has been the subject of numerous studies, which have created the modern field of Molecular Biology.

Central Dogma and Transcription

It is proposed that information flows from DNA to RNA in a process called transcription, and is then used to synthesize polypeptides by a process called translation. Transcription takes place in a manner similar to DNA replication.

A characteristic sequence of nucleotides marks the beginning of a gene on the DNA strand, and this region binds to a promoter protein that initiates RNA synthesis. The double stranded structure unwinds at the promoter site., and one of the strands serves as a template for RNA formation, as depicted in the following diagram. The RNA molecule thus formed is single stranded, and serves to carry information from DNA to the protein synthesis machinery called ribosomes. These RNA molecules are therefore called messenger-RNA (mRNA).

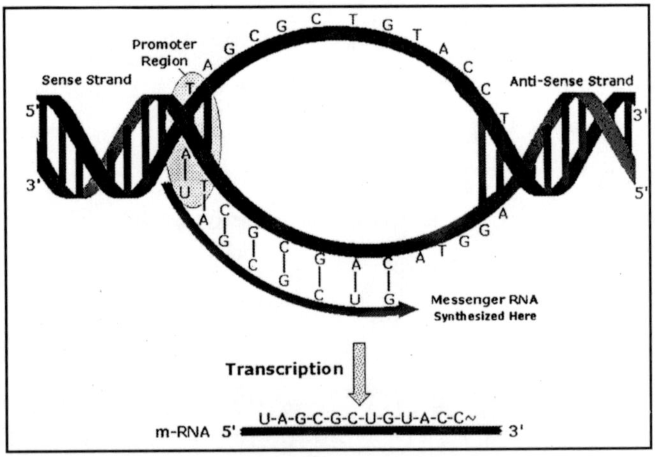

An important dstinction must be made here. One of the DNA strands in the double helix holds the genetic information used for protein synthesis. This is called the sense strand, or information strand. The complementary strand that binds to the sense strand is called the anti-sense strand and it seres as a template for generating a mRNA molecule that delivers a copy of the sense strand information to a ribosome. The promoter protein binds to a specific nuceotide sequence that identifies the sense strand, relative to the anti-sense strand. RNA synthesis is then initiated in the 3' direction, as nucleotide triphosphates bind to complementary bases on the template strand, and are joined by phosphate diester linkages. An animation of this process for DNA replication was presented earlier. A characteristic "stop sequence" of nucleotides terminates the RNA synthesis. The messenger molecule (coloured orange above) is released into the cytoplasm to find a ribosome, and the DNA then rewinds to its double helix structure.

In eucaryotic cells the initially transcribed m-RNA molecule is usually modified and shortened by an "editing" process that removes irrelevent material. The DNA of such organisms is often thousands of times larger and more complex than that composing the single chromosome of a procaryotic bacterial cell. This difference is due in part to repetitive nucleotide sequences (ca. 25% iu the human genome). Furthermore, over 95% of human DNA is found in intervening sequences that separate genes and parts of genes. The informational DNA segments that make up genes are called exons, and the noncoding segments are called introns.

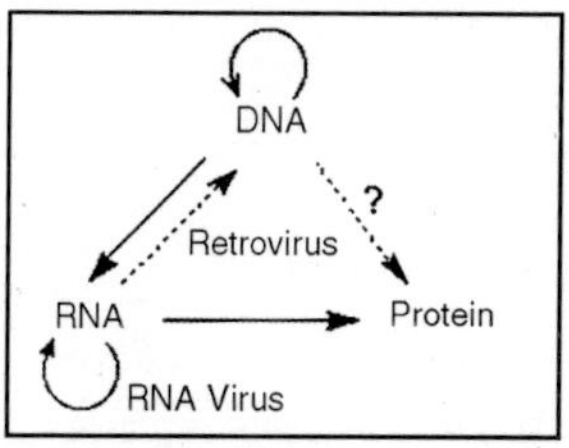

Before the mRNA molecule leaves the nucleus, the nonsense bases that make up the introns are cut out, and the informationally useful exons are joined together in a step known as RNA splicing. In this fashion shorter mRNA molecules carrying the blueprint for a specific protein are sent on their way to the ribosome factories.

The Central Dogma of molecular biology, which at first was formulated as a simple linear progression of information from DNA to RNA to Protein, is summarized in the following illustration. The replication process on the left consists of passing information from a parent DNA molecule to daughter molecules. The middle transcription process copies this information to a mRNA molecule. Finally, this information is used by the chemical machinery of the ribosome to make polypeptides.

As more has been learned about these relationships, the central dogma has been refined to the representation displayed on the right. The dark blue arrows show the general, well demonstrated, information transfers noted above. It is now known that an RNA-dependent DNA polymerase enzyme, known as a reverse transcriptase, is able to transcribe a single-stranded RNA sequence into double-stranded DNA (magenta arrow).

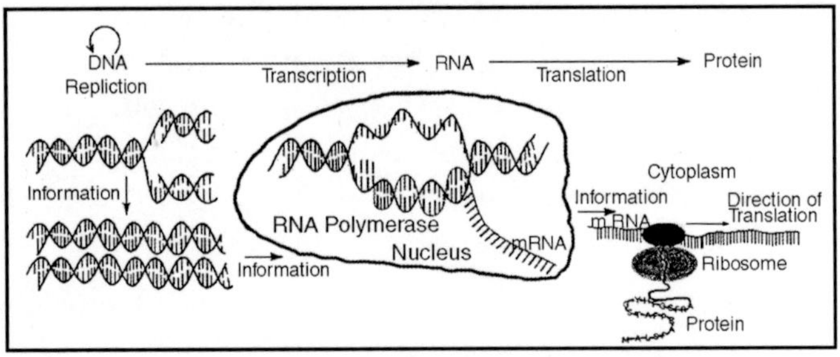

Such enzymes are found in all cells and are an essential component of retroviruses (*e.g.* HIV), which require RNA replication of their genomes (green arrow). Direct translation of DNA information into protein synthesis (orange arrow) has not yet been observed in a living organism. Finally, proteins appear to be an informational dead end, and do not provide a structural blueprint for either RNA or DNA. In the following section the last fundamental relationship, that of structural information translation from mRNA to protein, will be described

Translation

Translation is a more complex process than transcription. This would, of course, be expected. After all, the coded messages produced by the Enigma machine could be copied easily, but required a considerable decoding effort before they could be read with understanding.

In a similar sense, DNA replication is simply a complementary base pairing exercise, but the translation of the four letter (bases) alphabet code of RNA to the twenty letter (amino acids) alphabet of protein literature is far from trivial. Clearly, there could not be a direct one-to-one correlation of bases to amino acids, so the nucleotide letters must form short words or codons that define specific amino acids.

Many questions pertaining to this genetic code were posed:

- How many RNA nucleotide bases designate a specific amino acid?

  If separate groups of nucleotides, called codons, serve this purpose, at least three are needed. There are $4^3 = 64$ different nucleotide triplets, compared with $4^2 = 16$ possible pairs.

- Are the codons linked separately or do they overlap?

  Sequentially joined triplet codons will result in a nucleotide chain three times longer than the protein it describes. If overlapping codons are used then fewer total nucleotides would be required.

- If triplet segments of mRNA designate specific amino acids in the protein, how are the codons identified?

For the sequence ~CUAGGU~ are the codons CUA & GGU or ~C, UAG & GU~ or ~CU, AGG & U~?

- Are all the codon words the same size?

  In Morse code the most widely used letters are shorter than less common letters. Perhaps nature employs a similar scheme.

Physicists and mathematicians, as well as chemists and microbiologists all contributed to unravelling the genetic code. Although earlier proposals assumed efficient relationships that correlated the nucleotide codons uniquely with the twenty fundamental amino acids, it is now apparent that there is considerable redundancy in the code as it now operates. Furthermore, the code consists exclusively of non-overlapping triplet codons. Clever experiments provided some of the earliest breaks in deciphering the genetic code. It is found that RNA from many different organisms could initiate specific protein synthesis when combined with broken E.coli cells (the enzymes remain active). A synthetic polyuridine RNA induced synthesis of poly-phenylalanine, so the UUU codon designated phenylalanine. Likewise an alternating ~CACA~ RNA led to synthesis of a ~His-Thr-His-Thr~ polypeptide. The following table presents the present day interpretation of the genetic code.

		Second Position					
		U	C	A	G		
First Position	U	UUU Phe [F]	UCU Ser [S]	UAU Tyr [Y]	UGU Cys [C]	U	Third Position
		UUC Phe [F]	UCC Ser [S]	UAC Tyr [Y]	UGC Cys [C]	C	
		UUA Leu [L]	UCA Ser [S]	UAA Stop	UGA Stop	A	
		UUG Leu [L]	UCG Ser [S]	UAG Stop	UGG Trp [W]	G	
	C	CUU Leu [L]	CCU Pro [P]	CAU His [H]	CGU Arg [R]	U	
		CUC Leu [L]	CCC Pro [P]	CAC His [H]	CGC Arg [R]	C	
		CUA Leu [L]	CCA Pro [P]	CAA Gln [Q]	CGA Arg [R]	A	
		CUG Leu [L]	CCG Pro [P]	CAG Gln [Q]	CGG Arg [R]	G	
	A	AUU Ile [I]	ACU Thr [T]	AAU Asn [N]	AGU Ser [S]	U	
		AUC Ile [I]	ACC Thr [T]	AAC Asn [N]	AGC Ser [S]	C	
		AUA Ile [I]	ACA Thr [T]	AAA Lys [K]	AGA Arg [R]	A	
		AUG Met [M]	ACG Thr [T]	AAG Lys [K]	AGG Arg [R]	G	
	G	GUU Val [V]	GCU Ala [A]	GAU Asp [D]	GGU Gly [G]	U	
		GUC Val [V]	GCC Ala [A]	GAC Asp [D]	GGC Gly [G]	C	
		GUA Val [V]	GCA Ala [A]	GAA Glu [E]	GGA Gly [G]	A	
		GUG Val [V]	GCG Ala [A]	GAG Glu [E]	GGG Gly [G]	G	

Note that this is the RNA alphabet, and an equivalent DNA codon table would have all the U nucleotides replaced by T. Methionine and tryptophan are uniquely represented by a single codon. At the other extreme, leucine is represented by eight codons. The average redundancy for the twenty amino acids is about three. Also, there are three stop codons that terminate polypeptide synthesis.

RNA Codons for Protein Synthesis

The translation process is fundamentally straightforward. The mRNA strand bearing the transcribed code for synthesis of a protein interacts with relatively small RNA molecules (about 70-nucleotides) to which individual amino acids have been attached by an ester bond at the 3'-end.

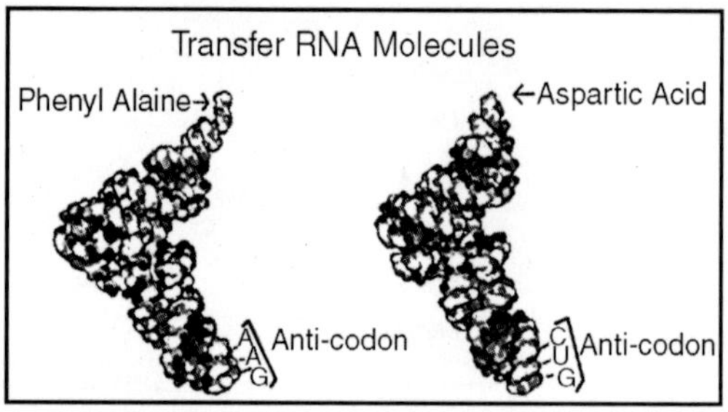

These transfer RNA's(tRNA) have distinctive three-dimensional structures cosisting of loops of single-stranded RNA connected by double stranded segments. This cloerleaf secondary structure is further wrapped into an "L-shaped" assembly, having the amino acid at the end of one arm, and a characteristic anti-codon region at the other end.

The anti-codon consists of a nucleotide triplet that is the complement of the amino acid's codon(s). Models of two such tRNA molecules are shown to the right. When read from the top to the bottom, the anti-codons depicted here should complement a codon in the previous table. A cell's protein synthesis takes place in organelles called ribosomes. Ribosomes are complex structures made up of two distinct and separable subunits (one about twice the size of the other). Each subunit is composed of one or two RNA molecules (60-70%) associated with 20 to 40 small proteins (30-40%). The ribosome accepts a mRNA molecule, binding initially to a characteristic nucleotide sequence at the 5'-end (coloured light blue in the following diagram).

This unique binding assures that polypeptide synthesis starts at the right codon. AtRNA molecule with the appropriate ant-codon thn attaches at the starting point and this is followed by a series of adjacent tRNA attachments, peptide bond formation and shifts of the ribosome along the mRNA chain to expose new codons to the ribosomal chemistry.

The following diagram is designed as a slide show illustrating these steps. The outcome is snthesis of a polypeptide chain corresponding to the mRNA blueprint. A "stop codon" at a designated position on the mRNA terminates the synthesis by introduction of a "Release Factor".

Drug acting on DNA

Mechanism of Intercalating

Intercalating agents wedge between bases along the DNA. The intercalated drug molecules affect the structure of the DNA, preventing polymerases and other DNA binding proteins from functioning properly. The result is prevention of DNA synthesis, inhibition of transcription and induction of mutations. The effect of principal alkaloids (sanguinarine, chelerythrine, coptisine, chelidonine) of greater celandine Chelidonium majus L., as well as the alkaloids from Colchicum autumnale L.(colchicine and colchamine) on calcium accumulation and oxidative phosphorylation in rat liver mitochondria has been studied.

The obtained data were compared with DNA intercalating properties of alkaloids detected by the method of thermodenaturation (DNA melting curve plots). It was found that chelerythrin and sanguinarine blocked absorption and accumulation of calcium cations and inhibited oxidative phosphorylation, while the coptisine significantly diminished those indices. Chelidonine, colchicines and colchamine had no influence on the studied characteristics. The effect of alkaloids upon mitochondria functional state correlated tightly with their DNA intercalating properties: chelerythrine and sanguinarine were strong intercalators, while coptisine was a weak one, and chelidonine, colchicine and colchamine did not interact with DNA and caused no changes in its melting point.

Correlation coefficient between the intercalating properties of alkaloids and their inhibition of calcium accumulation was.89, and with their oxidative phosphorylation inhibition -.93. It is suggested that the effect of studied alkaloids upon functional properties of mitochondria can be mediated by mtDNA. DNA topoisomerase I is essential for cellular metabolism and survival.

It is also the target of a novel class of anticancer drugs active against previously refractory solid tumors, the camptothecins. The present review describes the topoisomerase I catalytic mechanisms with particular emphasis on the cleavage complex that represents the enzyme's catalytic intermediate and the site of action for camptothecins. Roles of topoisomerase I in DNA replication, transcription and recombination are also reviewed. Because of the importance of topoisomerase I as a chemotherapeutic target, we review the mechanisms of action of camptothecins and the other topoisomerase I inhibitors identified to date.

The DNA intercalating agents 4'-(9-acridinyl-amino) methanesulphon-m-anisidide (m-AMSA) and adriamycin were studied by using filter elution methods to measure DNA single-strand breaks (SSB's), DNA-protein cross-links (DPC's), and double-stranded breaks (DSB's) in mouse leukemia L1210 cells. Both compounds produced SSB's and DPC's at nearly 1:1 ratios. The SSB's and DPC's were shown to be localized with respect to each other; this was inferred from the finding that filter assays based on protein adsorption completely prevented the elution of the DNA single-strand segments between SSB's. In the case of m-AMSA, which produces relatively high frequencies of DNA lesions, the possibility that a protein bridges across the SSB was excluded by alkaline sedimentation studies.

The o-AMSA isomer is much less cytotoxic than m-AMSA and did not produce protein-associated strand breaks. The simplest model to explain the results is that a protein becomes covalently bound to either the 3' or the 5' termini of the intercalator-induced strand breaks.

At moderately cytotoxic doses, m-AMSA yielded much larger frequencies of protein-associated SSB's than did adriamycin. m-AMSA-induced protein-associated SSB's saturated at approximately 60000 per cell over a concentration range in which m-AMSA uptake by the cells was proportional to the drug concentration. m-AMSA was found to enter and exit from cells very rapidly at 37 degrees C; protein-associated SSB's and DSB's also appeared and disappeared rapidly. At reduced temperature, however, the appearance and disappearance of protein-associated SSB's could be blocked while m-AMSA entry and exit still occurred.

The saturation behaviour and temperature dependence suggest that the formation and disappearance of protein-associated strand breaks is enzymatic. The simplest hypothesis is that the linked protein is a nuclease, such as a topoisomerase, which becomes bound to one terminus of the strand break it produces.

It is proposed that topoisomerases producing SSB's and DSB's are stimulated to different degrees by different intercalators. Topoisomerase II mediated DNA scission induced by both a nonintercalating agnt [4'-demethylepipodophyllotoxin 4-(4,6-O-ethylidene-beta-D-glucopyranoside)(VP-16)] and an intercalator [4'-(9-acridinylamino) methanesulphon-m-anisidide (m-AMSA)] wasstudied as a function of proliferation in Chinese hamster ovary (CHO), HeLa, an mouse leukemia L1210 cell lines. Log-phase CHO cells exhibited dose-dependent drug-induced DNA breaks, while plateau cells were found to be resistant to the effects of VP-16 and m-AMSA. Neither decreased viability nor altered drug uptake accounted for the drug resistance of these confluent cells. In contrast to CHO cells, plateau-phase HeLa and L1210 cells remained sensitive to VP-16 and m-AMSA. Recovery of drug sensitivity by plateau-phase CHO cells was found to reach a maximum approximately 18 h after these cells regained exponential growth and was independent of DNA synthesis.

DNA strand break frequency correlated with cytotoxicity in CHO cells; log cells demonstrated an inverse log linear relationship between drug dose (or DNA damage) and colony survival, whereas plateau-derived colony survival was virtually unaffected by increasing drug dose. Topoisomerase II activity, whether determined by decatenation of kinetoplast DNA, by cleavage of pBR322 DNA, or by precipitation of the DNA-topoisomerase II complex, was uniformly severalfold greater in log-phase CHO cells compared to plateau-phase cells.

The biochemical characteristics of the formation and disappearance of intercalator-induced DNA double-strand breaks (DSB) were studied in nuclei from mouse leukemia L1210 cells by using filter elution methodology [Bradley, M. O.,& Kohn, K.W.(1979) Nucleic Acids Res. 7, 793-804]. The three intercalators used were 4'-(9-acridinylamino)-methanesulphon-m-anisidide (m-AMSA), 5-iminodaunorubicin (5-ID), and ellipticine. These compounds differ in that they produced predominantly DNA single-strand breaks (SSB)(m-AMSA) or predominantly DNA double-strand breaks (ellipticine) or a mixture of both SSB and DSB (5-ID) in whole cells. In isolated nuclei, each intercalator produced DSB at a frequency comparable to that which is produced in whole cells. Moreover, these DNA breaks reversed within 30 min after drug removal. It thus appeared that neither ATP nor other nucleotides were necessary for intercalator-dependent DNA nicking-closing reactions. The formation of the intercalator-induced DSB was reduced at ice temperature. Break formation was also reduced in the absence of magnesium, at a pH above 6.4 and at NaCl concentrations above 200 mM. In the presence of ATP and AP analogues, the intercalator-induced cleavage was enhanced. These results suggest that the intercalator-induced DSB are enzymatically mediated and that the enzymes involved in these reactions can catalyze DNA double-strand cleavage and rejoining in the absence of ATP, although the occupancy of an ATP binding site might convert the enzyme to a form more reactive to intercalators. Three inhibitors of DNA topoisomerase II—novobiocin, nalidixic acid, and norfloxacin—reduced the formation of DNA strand breaks.

Mechanisms of Alkylating Aents

Alkylating agents work by three different mechanisms all of which achieve the same end result - disruption of DNA function and cell death.In the first mechanism an alkylating agent (represented in the figure below as a pink star) attaches alkyl groups (small carbon compounds-depicted as pink triangles) to

DNA bases. This alteration results in the DNA being fragmented by repair enzymes in their attempts to replace the alkylated bases (frame 3 of the diagram below). Alkylated bases prevent DNA synthesis and RN transcription from the affected DNA.

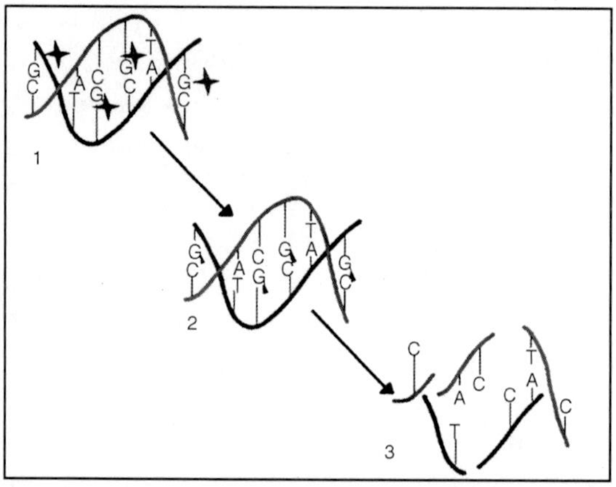

A second mechanism by which alkylating agents cause DNA damage is the formation of cross-bridges, bonds between atoms in the DNA (pink linkages below). In this process, two bases are linked together by an alkylating agent that has two DNA binding sites. Bridges can be formed within a single molecule of DNA (as shown below) or a cross-bridge may connect two different DNA molecules. Cross-linking prevents DNA from being separated for synthesis or transcription.

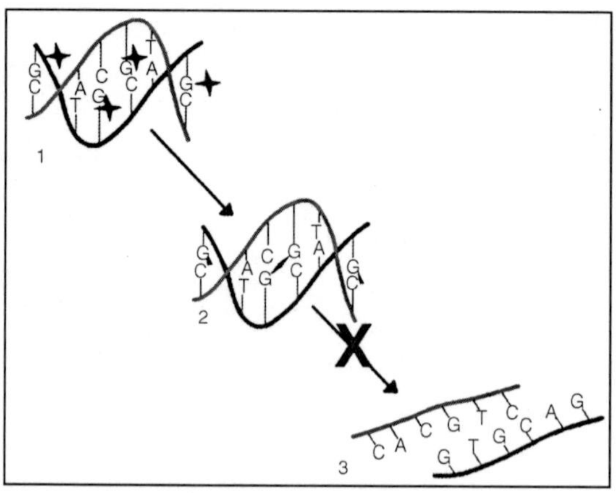

The third mechanism of action of alkylating agents is the induction of mispairing of the nucleotidesleading to mutations. In a normal DNA double helix, A always pairs with (is across from) T and G always pairs with C. As the figure below shows, alkylated G bases may erroneously pair with Ts. If this altered pairing is not corrected it may lead to a permanent mutation.

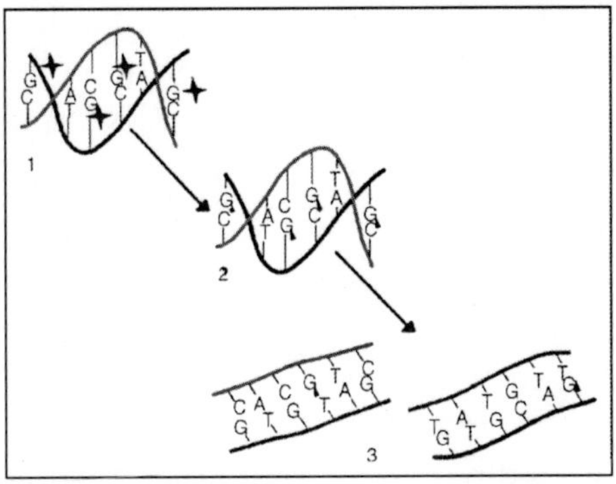

Alkylating agents are one of the earliest and most commonly used chemotherapy agents used for cancer treatments. Their use in cancer treatments started in early 1940s. Majority of alkaline agents are active or dormant nitrogen mustards, which are poisonous compound initially used for certain military purposes. Chlorambucil, Cyclophosphamide, CCNU, Melphalan, Procarbazine, Thiotepa, BCNU, and Busulphan are some of the commonly used alkylating agents.

Although they might differ in their clinical activity, action mechanism of all alkylating agents is the same. These agents work directly on the DNA and prevent the cell division process by cross-linking and breaking the DNA strands and causing abnormal base pairing. When a DNA is altered in this manner, undesired cellular activity comes to a halt and the cell dies eventually. Alkylating chemotherapy drugs are effective during all phases of cell cycle. Therefore, they are used to treat a larg number of cancers. However, they are more effective in treating slow-growing cancers such as solid tumors and leukemia. Long term use of alkylating agents can lead to permanent infertility by decreasing sperm production in males, and causing menstruation cessation in females. Many alkylating agents can also lead to secondary cancers such as Acute Myeloid Leukemia, years after the therapy.

Antimetabolites

Structure of antimetabolites (antineoplastic agents) is similar to certain compounds such as vitamins, amino acids, and precurors of DNA or RNA, found naturally in human body. Antimetabolites help in treatment cancer by inhibiting cell division thereby hindering the growth of tumor cells. These agents get incorporated in the DNA or RNA to interfere with the process of division of cancer cells.

Antimetabolites were first discovered and it is found that folic acid analog can reduce childhood leukemia. Out of 16 patients he tested, 10 displayed hematologic improvement. This discovery laid the foundation that enabled scientist to synthesize many new agents that could inhibit biological enzymatic reactions. Antimetabolites are found to be useful in treating chronic and acute cases of leukemia and various tumors. They are commonly used to treat gastrointestinal tract, breast, and ovary tumors.

Methotraxate, which is a commonly used antimetabolites chemotherapy agent, is effective in the S-phase of the cell cycle. It works by inhibiting an enzyme that is essential for DNA synthesis. 6-mercaptopurine and 5-fluorouracil (5FU) are two other commonly used antimetabolites. 5-Fluorouracil (5-FU) works by interfering with the DNA components, nucleotide, to stop DNA synthesis. This drug is used to treat many different types of cancers including breast, esophageal, head, neck, and gastric cancers. 6-mercaptopurine is an analogue of hypoxanthine and is commonly used to treat Acute Lymphoblastic Leukemia (ALL). Other popular antimetabolite chemotherapy drugs are Thioguanine, Cytarabine, Cladribine. Gemcitabine, and Fludarabine.

Anthracyclines

Anthracyclines were developed between 1970s and 1990s and are daunosamine and tetra-hydronaphthacenedione-based chemotherapy agents. These compounds are cell-cycle nonspecific and are used to treat a large number of cancers including lymphomas, leukemia, and uterine, ovarian, lung and breast cancers.

Anthracyclines drugs are developed from natural resources. For instance, daunorubicin is developed by isolating it from soil-dwelling fungus Streptomyces. Similarly, Doxorubicin, which is another commonly used anthracycline chemotherapy agent, is isolated from mutated strain of Streptomyces. Although both the drugs have similar clinical action mechanisms, doxorubicin is more effective in treating solid tumors. Idarubicin, Epirubicin, and Mitoxantrone are few of the other commonly used anthracycline chemotherapy drugs. Anthracyclines work by forming free oxygen radicals that breaks DNA strands thereby inhibiting DNA synthesis and function. These chemotherapeutic agents form a complex with DNA and enzyme to inhibit the topoisomerase enzyme. Topoisomerase is an enzyme class that causes the supercoiling of DNA, allowing DNA repair, transcription, and replication.

One of the main side effects of anthracyclines is that it can damage cells of heart muscle along with the DNA of cancer cell leading to cardiac toxicity.

Antitumor Antibiotics

Antitumor antibiotics are also developed from the soil fungus Streptomyces. These drugs are widely used to treat and suppress development of tumors in the body. Similar to anthracyclines, antitumor antibiotics drugs also form free oxygen radicals that result in DNA strand breaks, killing the growth of cancer cells. In most of the cases, these drugs are used in combination with other chemotherapy agents. Bleomycin is one of the commonly used antitumor antibiotic used to treat testicular cancer and hodgkin's lymphoma. The most serious side effect of this drug is lung toxicity that occurs when the oxygen radical formed by the antitumor antibiotics damages lung cells along with the cancer cells.

Monoclonal Antibodies

Monoclonal antibodies are one of the newer chemotherapy agents approved for cancer treatment by the Food and Drug Administration (FDA) in 1997. Alemtuzumab (Campath), Bevacizumab (Avastin), Cetuximab (Erbitux), Gemtuzumab (Mylotarg), Ibritumomab (Zevalin), Panitumumab (Vectibix), Rituximab (Rituxan), Tositumomab (Bexxar), and Trastuzumab (Herceptin) are some of the FDA approved monoclonal drugs used in chemotherapeutic cancer treatments. The treatment is known to be useful in treating colon, lung, head, neck, and breast cancers. Some of the monoclonal drugs are used to treat chronic lymphocytic leukemia, acute myelogenous leukemia, and non-Hodgkin's lymphoma.

Monoclonal antibodies work by attaching to certain parts of the tumor-specific antigens and make them easily recognizable by the host's immune system. They also prevent growth of cancer cells by blocking the cell receptors to which chemicals called 'growth factors' attach promoting cell growth.

Monoclonal antibodies can be combined with radioactive particles and other powerful anticancer drugs to deliver them directly to cancer cells. Using this method, long term radioactive treatment and anticancer drugs can be given to patients without causing any serious harm to other healthy cells of the body.

Platinum

Platinum-based natural metal derivatives were found to be useful for cancer treatments around 150 years ago with the synthesis of cisplatin. However, there clinical use did not commence until 30 years ago. Platinum-based chemotherapy agents work by cross-linking subunits of DNA. These agents act during any part of cell cycle and help in treating cancer by impairing DNA synthesis, transcription, and function.

Cisplatin, although found to be useful in treating testicular and lung cancer, is highly toxic and can severely damage the kidneys of the patient. Second generation platinum-complex carboplatin is found to be much less toxic in comparison to cisplatin and has fewer kidney-related side effects. Oxaliplatin, which is third generation platinum-based complex, is found to be helpful in treating colon cancer. Although, oxaliplatin does not cause any toxicity in kidney it can lead to severe neuropathies.

Plant Alkaloid

Plant alkaloid chemotherapy agents, as the name suggests, are plant derivatives. They are cell-specific chemotherapy agents. However, the cycle affected is based on the drug used for the treatment.

They are primarily categorized into four groups: topoisomerase inhibitors, vinca alkaloids, taxanes, and epipodophyllotoxins. Plant alkaloids are cell-cycle specific, but the cycle affected varies from drug to drug. Vincristine (Oncovin) is a plant alkaloid of interest in mesothelioma treatment.

Topoisomerase Inhibitors

Topoisomerase inhibitors are chemotherapy agents are categorized into Type I and Type II Topoisomerases inhibitors and they work by interfering with DNA transcription, replication, and function to prevent DNA supercoiling.

- *Type I Topoisomerase inhibitors:* These chemotherapy agents are extracted from the bark and wood of the Chinese tree Camptotheca accuminata. They work by forming a complex with topoisomerase DNA. This in turn suppresses the function of topoisomerase.

  Camptothecins which includes irinotecan and topotecan are commonly used type I topoisomerase inhibitors, first discovered in the late 1950s.

- *Type II Topoisomerase inhibitors:* These are extracted from the alkaloids found in the roots of May Apple plants. They work in the in the work in the late S and G2 phases of the cell cycle.

Amsacrine, etoposide, etoposide phosphate, and teniposide are some of the examples of type II topoisomerase inhibitors.

Vinca Alkaloids

Vinca alkaloids are derived from the periwinkle plant, Vinca rosea (Catharanthus roseus) and are known to be used by the natives of Madagascar to treat diabetes.

Although not useful in controlling diabetes, vinca alkaloids, are useful in treating leukemias.

They are effective in the M phase of the cell cycle and work by inhibiting tubulin assembly in microtubules. Vincristine, Vinblastine, Vinorelbine, and Vindesine are some of the popularly used vinca alkaloid chemotherapy agents used today. Major side effect of vinca alkaloids is that they can cause neurotoxicity in patients.

Taxanes

Taxanes are plant alkaloids that were first developed in 1963 by isolating it from first isolated from the bark of the Pacific yew tree, Taxus brevifolia in 1963. Paclitaxel, which is the active components of taxanes was first discovered in 1971 and was made available for clinical use in the year 1993. Taxanes also work in the M-phase of the cell cycle and inhibit the function of microtubules by binding with them. Paclitaxel and docetaxel are commonly used taxanes. Taxanes chemotherapy agents are used to treat a large array of cancers including breast, ovarian, lung, head and neck, gastric, esophageal, prostrate and gastric cancers. The main side effect of taxanes is that they lower the blood counts in patients.

Anti-Cancer Drug

Introduction

The available anticancer drugs have distinct mechanisms of action which may vary in their effects on different types of normal and cancer cells. A single "cure" for cancer has proved elusive since there is not a single type of cancer but as many as 100 different types of cancer. In addition, there are very few demonstrable biochemical differences between cancerous cells and normal cells. For this reason the effectiveness of many anticancer drugs is limited by their toxicity to normal rapidly growing cells in the intestinal and bone marrow areas. A final problem is that cancerous cells which are initially suppressed by a specific drug may develop a resistance to that drug. For this reason cancer chemotherapy may consist of using several drugs in combination for varying lengths of time.

Cancer Chemotherapy

Chemotherapy drugs, are sometimes feared because of a patient's concern about toxic effects. Their role is to slow and hopefully halt the growth and spread of a cancer. There are three goals associated with the use of the most commonly-used anticancer agents.

- Damage the DNA of the affected cancer cells.
- Inhibit the synthesis of new DNA strands to stop the cell from replicating, because the replication of the cell is what allows the tumor to grow.
- Stop mitosis or the actual splitting of the original cell into two new cells. Stopping mitosis stops cell division (replication) of the cancer and may ultimately halt the progression of the cancer.

Unfortunately, the majority of drugs currently on the market are not specific, which leads to the many common side effects associated with cancer chemotherapy. Because the common approach of all chemotherapy is to decrease the growth rate (cell division) of the cancer cells, the side effects are seen in bodily systems that naturally have a rapid turnover of cells iincluding skin, hair, gastrointestinal, and bone marrow. These healthy, normal cells, also end up damaged by the chemotherapy programme.

Catagories of Chemotherapy Drugs

In general, chemotherapy agents can be divided into three main categories based on their mechanism of action.

Stop the synthesis of pre DNA molecule building blocks: These agents work in a number of different ways. DNA building blocks are folic acid, heterocyclic bases, and nucleotides, which are made naturally within cells. All of these agents work to block some step in the formation of nucleotides or deoxyribon-ucleotides (necessary for making DNA). When these steps are blocked, the nucleotides, which arethe building blocks of DNA and RNA, can not be synthesized. Thus the cells can not replicate because they can nnot make DNA without the nucleotides.

Examples of drugs in this class include:

- Methotrexate (Abitrexate®),
- Fluorouracil (Adrucil®),
- Hydroxyurea (Hydrea®), and
- Mercaptopurine (Purinethol®).

Directly damage the DNA in the nucleus of the cell:

These agents chemically damage DNA and RNA. They disrupt replication of the DNA and either totally halt replication or cause the manufacture of nonsense DNA or RNA (*i.e.* the new DNA or RNA does not code for anything useful).

Examples of drugs in this class include cisplatin (Platinol®) and antibiotics - daunorubicin (Cerubidine®), doxorubicin (Adriamycin®), and etoposide (VePesid®).

Effect the synthesis or breakdown of the mitotic spindles:

Mitotic spindles serve as molecular railroads with "North and South Poles" in the cell when a cell starts to divide itself into two new cells. These spindles are very important because they help to split the newly copied DNA such that a copy goes to each of the two new cells during cell division. These drugs disrupt the formation of these spindles and therefore interrupt cell division.

Examples of drugs in this class of miotic disrupters include: Vinblastine (Velban), Vincristine (Oncovin) and Pacitaxel (Taxol).

Methotrexate: Methotrexate inhibits folic acid reductase which is responsible for the conversion of folic acid to tetrahydrofolic acid. At two stages in the biosynthesis of purines (adenine and guanine) and at one stage in the synthesis of pyrimidines (thymine, cytosine, and uracil), one-carbon transfer reactions occur which require specific coenzymes synthesized in the cell from tetrahydrofolic acid. Tetrahydrofolic acid itself is synthesized in the cell from folic acid with the help of an enzyme, folic acid reductase. Methotrexate looks a lot like folic acid to the enzyme, so it binds to it thinking that it is folic acid. In fact, methotrexate looks so good to the enzyme that it binds to it quite strongly and inhibits the enzyme. Thus, DNA synthesis cannot proceed because the coenzymes needed for one-carbon transfer reactions are not produced from tetrahydrofolic acid because there is no tetrahydrofolic acid. Again, without DNA, no cell division.

5-Fluorouracil: 5-Fluorouracil (5-FU; Adrucil, Fluorouracil, Efudex, Fluoroplex) is an effective pyrimidine antimetabolite. Fluorouracil is synthesized into the nucleotide, 5-fluoro-2-deoxyuridine. This product acts as an antimetabolite by inhibiting the synthesis of 2-deoxythymidine because the carbon - fluorine bond is extremely stable and prevents the addition of a methyl group in the 5-position. The failure to synthesize the thymidine nucleotide results in little or no production of DNA.

Two other similar drugs include: gemcitabine (Gemzar) and arabinosylcytosine (araC). They all work through similar mechanisms.

Hydroxyurea: Hydroxyurea blocks an enzyme which converts the cytosine nucleotide into the deoxy derivative. In addition, DNA synthesis is further inhibited because hydroxyurea blocks the incorporation of the thymidine nucleotide into the DNA strand.

Mercaptopurine: Mercaptopurine, a chemical analog of the purine adenine, inhibits the biosynthesis of adenine nucleotides by acting as an antimetabolite. In the body, 6-MP is converted to the corresponding ribonucleotide. 6-MP ribonucleotide is a potent inhibitor of the conversion of a compound called inosinic acid to adenine Without adenine, DNA cannot be synthesized. 6-MP also works by being incorporated into nucleic acids as thioguanosine, rendering the resulting nucleic acids (DNA, RNA) unable to direct proper protein synthesis. Adenine and Mercaptopurine - Chime in new window

Thioguanine: Thioguanine is an antimetabolite in the synthesis of guanine nucleotides.

Alkylating Agents: Alkylating agents involve reactions with guanine in DNA. These drugs add methyl or other alkyl groups onto molecules where they do not belong. This in turn inhibits their correct utilization by base pairing and causes a miscoding of DNA.

In the first mechanism an alkylating agent attaches alkyl groups to DNA bases. This alteration results in the DNA being fragmented by repair enzymes in their attempts to replace the alkylated bases. A second mechanism by which alkylating agents cause DNA damage is the formation of cross-bridges, bonds between atoms in the DNA. In this process, two bases are linked together by an alkylating agent that has two DNA binding sites. Cross-linking prevents DNA from being separated for synthesis or transcription. The third mechanism of action of alkylating agents causes the mispairing of the nucleotides leading to mutations.

There are six groups of alkylating agents: nitrogen mustards; ethylenimes; alkylsulphonates; triazenes; piperazines; and nitrosureas. Cyclosporamide is a classical example of the role of the host metabolism in the activation of an alkylating agent and is one or the most widely used agents of this class. It was hoped that the cancer cells might posses enzymes capable of accomplishing the cleavage, thus resulting in the selective production of an activated nitrogen mustard in the malignant cells. Compare the top and bottom structures in the graphic on the left.

Antibiotics: A number of antibiotics such as anthracyclines, dactinomycin, bleomycin, adriamycin, mithramycin, bind to DNA and inactivate it. Thus the synthesis of RNA is prevented.

General properties of these drugs include: interaction with DNA in a variety of different ways including intercalation (squeezing between the base pairs), DNA strand breakage and inhibition with the enzyme topoisomerase II. Most of these compounds have been isolated from natural sources and antibiotics. However, they lack the specificity of the antimicrobial antibiotics and thus produce significant toxicity.

Th anthracyclines are among the most important antitumor drugs available. Doxorubicin is widely used for the treatment of several solid tumors while daunorubicin and idarubicin are used exclusively for the treatment of leukemia.

These agents have a number of important effects including: intercalating (squeezing between the base pairs) with DNA affecting many functions of the DNA including DNA and RNA synthesis. Breakage of the DNA strand can also occur by inhibition of the enzyme topoisomerase II.

Dactinomycin (Actinomycin D): At low concentrations dactinomycin inhibits DNA directed RNA synthesis and at higher concentrations DNA synthesis is also inhibited. All types of RNA are affected, but ribosomal RNA is more sensitive. Dactinomycin binds to double stranded DNA, permitting RNA chain initiation but blocking chain elongation. Binding to the DNA depends on the presence of guanine.

Mitotic Disrupters: Plant alkaloids like vincristine prevent cell division, or mitosis. There are several phases of mitosis, one of which is the metaphase. During metaphase, the cell pulls duplicated DNA chromosomes to either side of the parent cell in structures called "spindles". These spindles ensure that each new cell gets a full set of DNA. Spindles are microtubular fibres formed with the help of the protein "tubulin". Vincristine binds to tubulin, thus preventing the formation of spindles and cell division. In contrast to other microtubule antagonists, taxol disrupts the equilibrium between free tubulin and mircrotubules by shifting it in the direction of assembly, rather than disassembly. As a result, taxol treatment causes both the stabilization of microtubules and the formation of abnormal bundles of microtubules. The net effect is still the disruption of mitosis.

Ribonucleic Acid

Ribonucleic acid (RNA) is a biologically important type of molecule that consists of a long chain of nucleotide units. Each nucleotide consists of a nitrogenous base, a ribose sugar, and a phosphate. RNA is very similar to DNA, but differs in a few important structural details: in the cell, RNA is usually single-stranded, while DNA is usually double-stranded; RNA nucleotides contain ribose while DNA contains deoxyribose (a type of ribose that lacks one oxygen atom); and RNA has the base uracil rather than thymine that is present in DNA.

Ribonucleic acid (RNA) is a linear molecule composed of four types of smaller molecules called ribonucleotide bases: adenine (A), cytosine (C), guanine (G), and uracil (U). RNA is often compared to a copy from a reference book, or a template, because it carries the same information as its DNA template but is not used for long-term storage. Each ribonucleotide base consists of a ribose sugar, a phosphate group, and a nitrogenous base. Adjacent ribose nucleotide bases are chemically attached to one another in a chain via chemical bonds called phosphodiester bonds. Unlike DNA, RNA is usually single-stranded. Additionally, RNA contains ribose sugars rather than deoxyribose sugars, which makes RNA more unstable and more prone to degradation. RNA is synthesized from DNA by an enzyme known as RNA polymerase during a process called transcription.

The new RNA sequences are complementary to their DNA template, rather than being identical copies of the template. RNA is then translated into proteins by structures called ribosomes. There are three types of RNA involved in the translation process: messenger RNA (mRNA), transfer RNA (tRNA), and ribosomal RNA (rRNA). Although some RNA molecules are passive copies of DNA, many play crucial, active roles in the cell. For example, some RNA molecules are involved in switching genes on and off, and other RNA molecules make up the critical protein synthesis machinery in ribosomes.

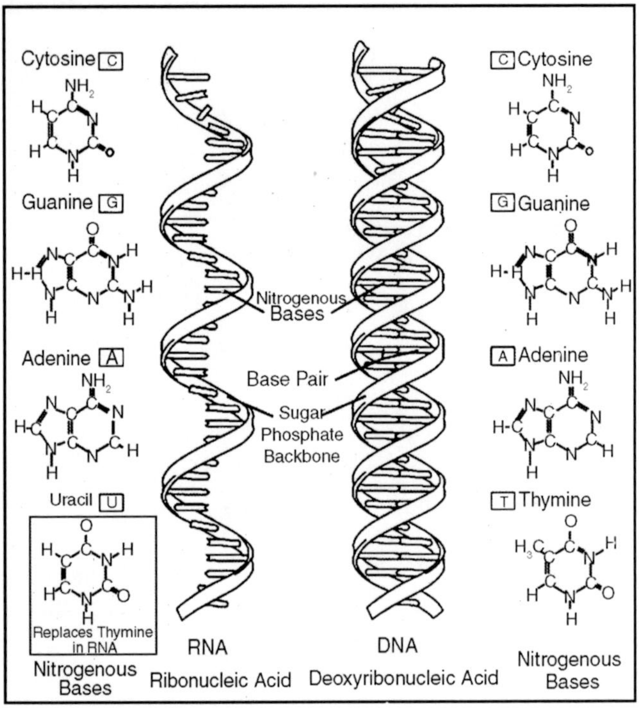

RNA is transcribed from DNA by enzymes called RNA polymerases and is generally further processed by other enzymes. RNA is central to the synthesis of proteins. Here, a type of RNA called messenger RNA carries information from DNA to structures called ribosomes.

These ribosomes are made from proteins and ribosomal RNAs, which come together to form a molecular machine that can read messenger RNAs and translate the information they carry into proteins. There are many RNAs with other roles – in particular regulating which genes are expressed, but also as the genomes of most viruses.

Messenger RNA (mRNA) and its Precursors

The primary RNA transcripts are produced by copying the DNA of a protein-encoding gene.[1] Subsequent processing steps include addition, removal, and modification of nucleotides as well as splicing events that excise internal segments. The mature mRNA is translated to produce a protein whose amino acid sequence is determined by the sequence of the coding region in the gene. The messenger RNA, as the name implies, is the molecule that carries the message from the gene to the protein synthesis machinery.

Ribosomal RNA (rRNA)

The ribosomes are the most important part of the translation machinery and it has long been known that much of the mass of ribosomes is due to the presence several types of ribosomal RNA. These are noncoding RNAs produced by transcription of ribosomal RNA genes.[3] One of the key steps in translation—formation of the peptide bond—is catalyzed by the rRNA component of the ribosome. It is the major catalytic RNA in cells.

Transfer RNA (tRNA)

tRNAs are intermediates in protein synthesis. There are many different tRNA molecules in every cell and each one binds a specific amino acid, yielding an aminoacylated-tRNA (aa-tRNA). Each different aminoacylated tRNA interacts with a particular codon in mRNA thus delivering the correct amino acid to to the site of protein synthesis.

Small RNAs

The small RNAs represent a heterogeneous category of RNAs covering a wide ranges of functions. Some of them have catalytic functions—RNAse P is the classic example. Some of them are structural components of ribonucleoprotein complexes (*e.g.* signal recognition particle). Some of them are guide RNAs involved in various processing events. The best known examples of guide RNAs are the small RNAs of the spliceosome complexes that mediate the splicing of mRNA precursors.[2]Other small RNAs were known to be involved in the regulation of gene expression.

Genomic RNA

Some viruses, notably retroviruses, have an RNA genome instead of a DNA genome. In addition, the mobility of various transposons is due to an intermediate RNA copy of the transposon sequence (retrotransposons). This was the state of knowledge 25 years ago. Since then, the study of RNA has made remarkable progress. Our knowledge of all the fundamental processes—transcription, processing, and catalysis—has expanded enormously.

The biggest change is in the area of small RNAs. Today there are several categories of small RNAs—siRNA, microRNA, piRNA—that were only discovered in the past 10-15 years. The functions of these small RNA molecules are still being worked out. There's little doubt that some of them have important biological roles but there's considerable controversy over what percentage might be artifacts of one sort or another.

RNA replaces DNA as the genetic material. As with DNA, RNA follows specific base pairing rules, except that in RNA the base uracil replaces the base thymine (*i.e.*, instead of an adenine-thymine or A-T pairing, there is an adenine-uracil or A-U pairing). Accordingly, when RNA acts as a carrier of genetic information, uracil replaces thymine in the genetic code. In humans, messenger RNA (mRNA) is the product of transcription and acts to convey genetic information from the nucleus to the protein assembly complex at the ribosome. The ribsome is composed of ribosomal RNA (rRNA) and other proteins. Transfer RNAs (tRNA) act to catalyze the translation process by acting as carriers of specific amino acids. Because tRNAs bind to specific sites on the strand of mRNA, the sequence of amino acids subsequently inserted into the synthesized protein is both specific and genetically determined by the nucleotide sequence in DNA from which the mRNA strand was originally transcribed.

Other forms of RNA perform important roles in other biochemical reactions. Regardless of function, RNA is a biopolymer made up of ribonucleotide units and is present in all living cells and some viruses. The chemical units of RNA are ribonucleotide monomers consisting of a ribose sugar ($C_5H_{10}O_5$) phosphorylated at the third carbon (C3) and linked to one of four bases through a type of chemical linkage formed between a sugar and a base by a condensation reaction (glycosidic bond). The four bases found in RNA are adenine (A), guanine (G), cytosine (C), and uracil (U). Other bases may also be found, although they are generally modified versions of these four (*e.g.*, methylated bases are found in parts of tRNA).

The single nucleotides (monomers) of RNA form a linear chain by linking their phosphate groups and sugars in phosphodiester bonds. RNA does not form a double stranded alpha-helix as does DNA. In some parts of the RNA molecule, there is folding into alpha-helical-like regions. Corresponding to their unique functions, messenger RNA (mRNA), ribosomal RNA (rRNA), and transfer RNA (tRNA) all have different three-dimensional structures. In higher eukaryotic organisms, different RNAs are found distributed throughout the cell—in the nucleus, cytoplasm, and also in cytoplasmic organelles such as mitochondria and, in plants, chloroplasts.

The nucleus is the chief site of RNA synthesis and the source of all cytoplasmic RNA, while mitochondria and chloroplasts synthesize their RNA from their own DNA. rRNA is synthesized by the nucleoli within the nucleus, while the high molecular weight precursor to cytoplasmic mRNA, sometimes termed heterogeneous nuclear or hnRNA, is transcribed on the DNA chromatin. Low molecular weight RNA also occurs in the nucleus and consists partly of tRNA and partly of RNA, which has a regulatory function in geneactivation. The cytoplasm contains tRNA and rRNA in the ribosomes and mRNA in polysomes, or polyribosomes. The latter are the structural units of protein biosynthesis, consisting of several ribosomes attached to a strand of mRNA.

The function of mRNA is to transcribe the information held in DNA. In the cells of eukaryotic organisms, the first transcriptional product is the long, heterogenous nuclear RNA, or hnRNA. This contains both the nucleotide sequences eventually transcribed into polypeptides and large tracts of sequences not translated. Non-translated sequences are termed introns (or intervening sequences). Removal of introns, and other untranslated portions of the molecule, edits hnRNA into mRNA molecules. After editing removes as much as 90% of hnRNA, the resulting mRNA molecules are transported into the cytoplasm.

rRNA is located within ribosomes, the sites of protein biosynthesis. Ribosomes are large ellipsoid cytoplasmic organelles consisting of RNA and protein. tRNA, the smallest known functional RNA, is essential for protein biosynthesis. Its purpose is to transfer a specific amino acid from the cytoplasm and incorporate it into the growing polypeptide chain on the polysome. Different tRNAs contain between 70 and 85 nucleotides. The most characteristic feature of tRNA is that it contains the anticodon, a sequence of three nucleotides specific for the mRNA codon sequence. There is at least one tRNA per cell bearing the anticodon for each of the 20 amino acids. The aminoacyl-tRNA (the tRNA carrying the amino acid) binds to the large subunit of a ribosome, where antiparallel basepairing occurs between the anticodon of the tRNA and the complementary codon of the associated mRNA. The specificity of this base pairing ensures that the amino acid inserts into the correct position in the growing protein polypeptide chain. During translation, the deacylated tRNA (*i.e.*, with its amino acid removed) is released from the ribosome and becomes available once again for recharging with its amino acid.

DNA-dependent RNA synthesis is the process of RNA sythesis on a template of DNA. According to the rules of base pairing, the base sequence of DNA determines the synthesis of a complementary base sequence in RNA. Assisted (catalyzed) by the enzyme RNA polymerase, the growing RNA chain releases from the template so that the process can start again, even before the previous molecule is complete. Termination codons and a termination factor known as rho-factor end the synthesis process. In certain viruses, RNA-dependent RNA synthesis occurs, with the viral RNA acting as a template for the synthesis of new RNA.

Each nucleotide in RNA contains a ribose sugar, with carbons numbered 1' through 5'. A base is attached to the 1' position, generally adenine (A), cytosine (C), guanine (G) or uracil (U). Adenine and guanine are purines, cytosine and uracil are pyrimidines. A phosphate group is attached to the 3' position of one ribose and the 5' position of the next. The phosphate groups have a negative charge each at physiological pH, making RNA a charged molecule (polyanion). The bases may form hydrogen bonds between cytosine and guanine, between adenine and uracil and between guanine and uracil.However other interactions are possible, such as a group of adenine bases binding to each other in a bulge, or the GNRA tetraloop that has a guanine–adenine base-pair.

An important structural feature of RNA that distinguishes it from DNA is the presence of a hydroxyl group at the 2' position of the ribose sugar. The presence of this functional group causes the helix to adopt the A-form geometry rather than the B-form most commonly observed in DNA. This results in a very deep and narrow major groove and a shallow and wide minor groove. A second consequence of the presence of the 2'-hydroxyl group is that in conformationally flexible regions of an RNA molecule (that is, not involved in formation of a double helix), it can chemically attack the adjacent phosphodiester bond to cleave the backbone. RNA is transcribed with only four bases (adenine, cytosine, guanine and uracil), but there are numerous modified bases and sugars in mature RNAs. Pseudouridine (Ø), in which the linkage between uracil and ribose is changed from a C–N bond to a C–C bond, and ribothymidine, are found in various places (most notably in the TØC loop of tRNA).Another notable modified base is hypoxanthine, a deaminated adenine base whose nucleoside is called inosine.

Inosine plays a key role in the wobble hypothesis of the genetic code.There are nearly 100 other naturally occurring modified nucleosides,of which pseudouridine and nucleosides with 2'-O-methylribose are the most common.The specific roles of many of these modifications in RNA are not fully understood. However, it is notable that in ribosomal RNA, many of the post-transcriptional modifications occur in highly functional regions, such as the peptidyl transferase centre and the subunit interface, implying that they are important for normal function.

RNA and DNA are both nucleic acids, but differ in three main ways. First, unlike DNA which is double-stranded, RNA is a single-stranded molecule in most of its biological roles and has a much shorter chain of nucleotides. Second, while DNA contains deoxyribose, RNA contains ribose, (there is no hydroxyl group attached to the pentose ring in the 2' position in DNA). These hydroxyl groups make RNA less stable than DNA because it is more prone to hydrolysis. Third, the complementary base to adenine is not thymine, as it is in DNA, but rather uracil, which is an unmethylated form of thymine.Like DNA, most biologically active RNAs, including mRNA, tRNA, rRNA, snRNAs and other non-coding RNAs, contain self-complementary sequences that allow parts of the RNA to fold and pair with itself to form double helices. Structural analysis of these RNAs have revealed that they are highly structured.

5

Genetic Algorithms

A genetic algorithm (GA) is a search heuristic that mimics the process of natural evolution. This heuristic is routinely used to generate useful solutions to optimisation and search problems. Genetic algorithms belong to the larger class of evolutionary algorithms (EA), which generate solutions to optimisation problems using techniques inspired by natural evolution, such as inheritance, mutation, selection, and crossover.

Methodology

In a genetic algorithm, a population of strings (called chromosomes or the genotype of the genome), which encode candidate solutions (called individuals, creatures, or phenotypes) to an optimisation problem, evolves toward better solutions. Traditionally, solutions are represented in binary as strings of 0s and 1s, but other encodings are also possible.

The evolution usually starts from a population of randomly generated individuals and happens in generations. In each generation, the fitness of every individual in the population is evaluated, multiple individuals are stochastically selected from the current population (based on their fitness), and modified (recombined and possibly randomly mutated) to form a new population.

The new population is then used in the next iteration of the algorithm. Commonly, the algorithm terminates when either a maximum number of generations has been produced, or a satisfactory fitness level has been reached for the population. If the algorithm has terminated due to a maximum number of generations, a satisfactory solution may or may not have been reached.

Genetic algorithms find application in bioinformatics, phylogenetics, computational science, engineering, economics, chemistry, manufacturing, mathematics, physics and other fields.

A typical genetic algorithm requires:

Genetic Representation

Genetic representation is a way of representing solutions/individuals in evolutionary computation methods. Genetic representation can encode appearance, behaviour, physical qualities of individuals. Designing a good genetic representation that is expressive and evolvable is a hard problem in evolutionary computation. Difference in genetic representations is one of the major criteria drawing a line between known classes of evolutionary computation.

Genetic algorithms use linear binary representations. The most standard one is an array of bits. Arrays of other types and structures can be used in essentially the same way. The main property that makes these genetic representations convenient is that their parts are easily aligned due to their fixed size. This facilitates simple crossover operation. Variable length representations were also explored in Genetic algorithms, but crossover implementation is more complex in this case. Evolution strategy uses linear real-valued representations, e.g. an array of real values. It uses mostly gaussian mutation and blending/averaging crossover.

Genetic programming (GP) pioneered tree-like representations and developed genetic operators suitable for such representations. Tree-like representations are used in GP to represent and evolve functional programs with desired properties.

Human-based genetic algorithm (HBGA) offers a way to avoid solving hard representation problems by outsourcing all genetic operators to outside agents, in this case, humans. The algorithm has no need for knowledge of a particular fixed genetic representation as long as there are enough external agents capable of handling those representations, allowing for free-form and evolving genetic representations.

Fitness Function

A fitness function is a particular type of objective function that is used to summarise, as a singlefigure of merit, how close a given design solution is to achieving the set aims. In particular, in the fields of genetic programming and genetic algorithms, each design solution is represented as a string of numbers (referred to as a chromosome). After each round of testing, or simulation, the idea is to delete the 'n' worst design solutions, and to breed 'n' new ones from the best design solutions. Each design solution, therefore, needs to be awarded a figure of merit, to indicate how close it came to meeting the overall specification, and this is generated by applying the fitness function to the test, or simulation, results obtained from that solution.

The reason that genetic algorithms are not a lazy way of performing design work is precisely because of the effort involved in designing a workable fitness function. Even though it is no longer the human designer, but the computer, that comes up with the final design, it is the human designer who has to design the fitness function.

If this is designed wrongly, the algorithm will either converge on an inappropriate solution, or will have difficulty converging at all. Moreover, the fitness function must not only correlate closely with the designer's goal, it must also be computed quickly. Speed of execution is very important, as a typical genetic algorithm must be iterated many times in order to produce a usable result for a non-trivial problem. Fitness approximation may be appropriate, especially in the following cases:

- Fitness computation time of a single solution is extremely high
- Precise model for fitness computation is missing
- The fitness function is uncertain or noisy.

Two main classes of fitness functions exist: one where the fitness function does not change, as in optimising a fixed function or testing with a fixed set of test cases; and one where the fitness function is mutable, as in niche differentiation or co-evolving the set of test cases.

Another way of looking at fitness functions is in terms of a fitness landscape, which shows the fitness for each possible chromosome.

Definition of the fitness function is not straightforward in many cases and often is performed iteratively if the fittest solutions produced by GA are not what is desired. In some cases, it is very hard or impossible to come up even with a guess of what fitness function definition might be. Interactive genetic algorithms address this difficulty by outsourcing evaluation to external agents (normally humans). A standard representation of the solution is as an array of bits. Arrays of other types and structures can be used in essentially the same way.

The main property that makes these genetic representations convenient is that their parts are easily aligned due to their fixed size, which facilitates simple crossover operations. Variable length representations may also be used, but crossover implementation is more complex in this case. Tree-like representations are explored in genetic programming and graph-form representations are explored in evolutionary programming. The fitness function is defined over the genetic representation and measures the quality of the represented solution.

The fitness function is always problem dependent. For instance, in the knapsack problem one wants to maximize the total value of objects that can be put in a knapsack of some fixed capacity. A representation of a solution might be an array of bits, where each bit represents a different object, and the value of the bit (0 or 1) represents whether or not the object is in the knapsack. Not every such representation is valid, as the size of objects may exceed the capacity of the knapsack.

The fitness of the solution is the sum of values of all objects in the knapsack if the representation is valid, or 0 otherwise. In some problems, it is hard or even impossible to define the fitness expression; in these cases, interactive genetic algorithms are used. Once the genetic representation and the fitness function are defined, a GA proceeds to initialize a population of solutions (usually randomly) and then to improve it through repetitive application of the mutation, crossover, inversion and selection operators.

Initialization

Initially many individual solutions are (usually) randomly generated to form an initial population. The population size depends on the nature of the problem, but typically contains several hundreds or thousands of possible solutions. Traditionally, the population is generated randomly, allowing the entire range of possible solutions (the search space). Occasionally, the solutions may be "seeded" in areas where optimal solutions are likely to be found

Selection (Genetic Algorithm)

Selection is the stage of a genetic algorithm in which individual genomes are chosen from a population for later breeding (recombination or crossover). A generic selection procedure may be implemented as follows:

1. The fitness function is evaluated for each individual, providing fitness values, which are then normalized. Normalization means dividing the fitness value of each individual by the sum of all fitness values, so that the sum of all resulting fitness values equals 1.
2. The population is sorted by descending fitness values.
3. Accumulated normalized fitness values are computed (the accumulated fitness value of an individual is the sum of its own fitness value plus the fitness values of all the previous individuals). The accumulated fitness of the last individual should be 1 (otherwise something went wrong in the normalization step).

4. A random number R between 0 and 1 is chosen.
5. The selected individual is the first one whose accumulated normalized value is greater than R.

If this procedure is repeated until there are enough selected individuals, this selection method is called fitness proportionate selection or roulette-wheel selection. If instead of a single pointer spun multiple times, there are multiple, equally spaced pointers on a wheel that is spun once, it is called stochastic universal sampling. Repeatedly selecting the best individual of a randomly chosen subset is tournament selection. Taking the best half, third or another proportion of the individuals is truncation selection.

There are other selection algorithms that do not consider all individuals for selection, but only those with a fitness value that is higher than a given (arbitrary) constant. Other algorithms select from a restricted pool where only a certain percentage of the individuals are allowed, based on fitness value. Retaining the best individuals in a generation unchanged in the next generation, is called elitism orelitist selection. It is a successful (slight) variant of the general process of constructing a new population.

Crossover (Genetic Algorithm)

In genetic algorithms, crossover is a genetic operator used to vary the programming of a chromosome or chromosomes from one generation to the next. It is analogous to reproduction and biological crossover, upon which genetic algorithms are based. Cross over is a process of taking more than one parent solutions and producing a child solution from them. There are methods for selection of the chromosomes. Those are also given below.

Methods of Selection of Chromosomes for Crossover

- Roulette wheel selection: It is also known as fitness proportionate selection. The individual is selected on the basis of fitness. The probability of an individual to be selected increases with the fitness of the individual greater or less than its competitor's fitness.
- Boltzmann selection
- Tournament selection
- Rank selection
- Steady state selection.

Crossover Techniques

Many crossover techniques exist for organisms which use different data structures to store themselves.

One-point Crossover

A single crossover point on both parents' organism strings is selected. All data beyond that point in either organism string is swapped between the two parent organisms. The resulting organisms are the children:

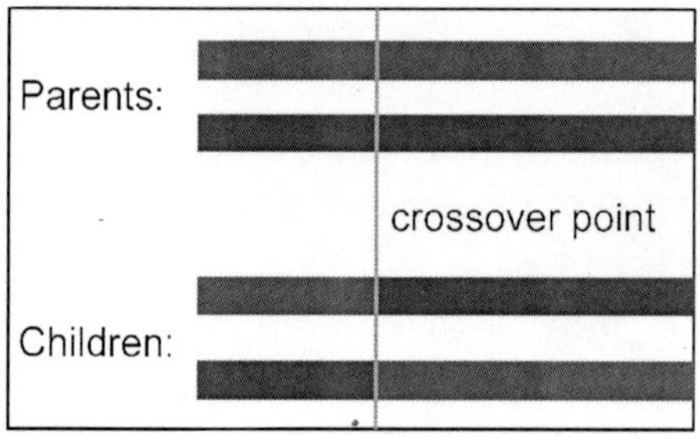

Two-point Crossover

Two-point crossover calls for two points to be selected on the parent organism strings. Everything between the two points is swapped between the parent organisms, rendering two child organisms:

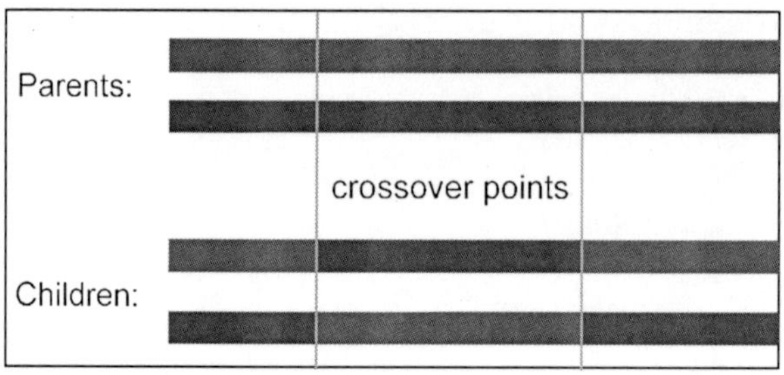

"Cut and Splice"

Another crossover variant, the "cut and splice" approach, results in a change in length of the children strings. The reason for this difference is that each parent string has a separate choice of crossover point.

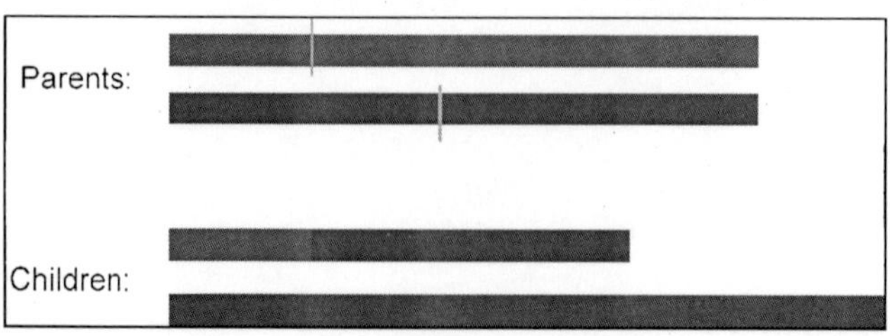

Uniform Crossover and Half Uniform Crossover

The Uniform Crossover uses a fixed mixing ratio between two parents. Unlike one- and two-point crossover, the Uniform Crossover enables the parent chromosomes to contribute the gene level rather than the segment level. If the mixing ratio is 0.5, the offspring has approximately half of the genes from first parent and the other half from second parent, although cross over points can be randomly chosen.

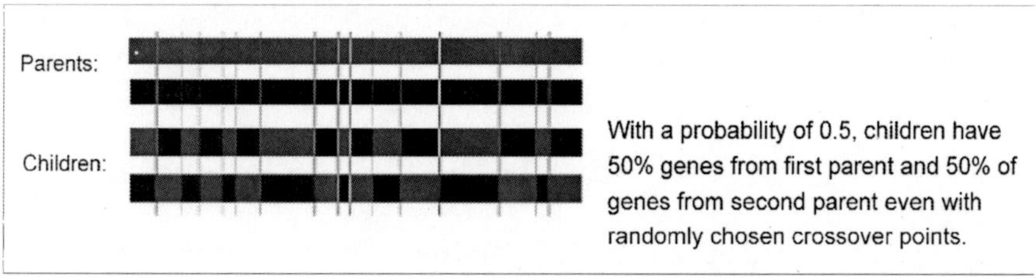

The Uniform Crossover evaluates each bit in the parent strings for exchange with a probability of 0.5. Even though the uniform crossover is a poor method, empirical evidence suggest that it is a more exploratory approach to crossover than the traditional exploitative approach that maintains longer schemata.

This results in a more complete search of the design space with maintaining the exchange of good information. Unfortunately, no satisfactory theory exists to explain the discrepancies between the Uniform Crossover and the traditional approaches.

In the uniform crossover scheme (UX) individual bits in the string are compared between two parents. The bits are swapped with a fixed probability, typically 0.2. In the half uniform crossover scheme (HUX), exactly half of the nonmatching bits are swapped. Thus first the Hamming distance (the number of differing bits) is calculated. This number is divided by two. The resulting number is how many of the bits that do not match between the two parents will be swapped.

Three Parent Crossover

In this technique, the child is derived from three parents. They are randomly chosen. Each bit of first parent is checked with bit of second parent whether they are same. If same then the bit is taken for the offspring otherwise the bit from the third parent is taken for the offspring.

parent1 1 1 0 1 0 0 0 1 0

parent2 0 1 1 0 0 1 0 0 1

parent3 1 1 0 1 1 0 1 0 1

offspring 1 1 0 1 0 0 0 0 1

Crossover for Ordered Chromosomes

Depending on how the chromosome represents the solution, a direct swap may not be possible. One such case is when the chromosome is an ordered list, such as an ordered list of the cities to be travelled for the travelling salesman problem. There are many crossover methods for ordered chromosomes. The already mentioned N-point crossover can be applied for ordered chromosomes also, but this always need a corresponding repair process, actually, some ordered crossover method are derived from the idea. However, sometimes a crossover of chromosomes produces recombinations which violate the constraint of ordering and thus need to be repaired.

Several examples for crossover operators (also mutation operator) preserving a given order are given in :

1. partially-matched crossover (PMX): In this method, two crossover points are selected at random and PMX proceeds by position wise exchanges. The two crossover points give matching selection. It affects cross by position-by-position exchange operations. In this method parents are mapped to each other hence we can also call it partially mapped crossover.
2. cycle crossover (CX): Beginning at any gene i in parent 1, the i-th gene in parent 2 becomes replaced by it. The same is repeated for the displaced gene until the gene which is equal to the first inserted gene becomes replaced (cycle).
3. order crossover operator (OX1): A portion of one parent is mapped to a portion of the other parent. From the replaced portion on, the rest is filled up by the remaining genes, where already present genes are omitted and the order is preserved.
4. order-based crossover operator (OX2)
5. position-based crossover operator (POS)
6. voting recombination crossover operator (VR)
7. alternating-position crossover operator (AP)
8. sequential constrictive crossover operator (SCX)

Other possible methods include the edge recombination operator and partially mapped crossover.

Crossover Biases

For crossover operators which exchange contiguous sections of the chromosomes (e.g. k-point) the ordering of the variables may become important. This is particularly true when good solutions contain building blocks which might be disrupted by a non-respectful crossover operator.

Mutation (Genetic Algorithm)

In genetic algorithms of computing, mutation is a genetic operator used to maintain genetic diversity from one generation of a population of algorithm chromosomes to the next. It is analogous to biological mutation. Mutation alters one or more gene values in a chromosome from its initial state. In mutation, the solution may change entirely from the previous solution. Hence GA can come to better solution by using mutation, mutation occurs during evolution according to a user-definable mutation probability. This probability should be set low. If it is set to high, the search will turn into a primitive random search.

The classic example of a mutation operator involves a probability that an arbitrary bit in a genetic sequence will be changed from its original state. A common method of implementing the mutation operator involves generating a random variable for each bit in a sequence. This random variable tells whether or not a particular bit will be modified. This mutation procedure, based on the biological point mutation, is called single point mutation. Other types are inversion and floating point mutation. When the gene encoding is restrictive as in permutation problems, mutations are swaps, inversions and scrambles.

The purpose of mutation in GAs is preserving and introducing diversity. Mutation should allow the algorithm to avoid local minima by preventing the population of chromosomes from becoming too similar to each other, thus slowing or even stopping evolution. This reasoning also explains the fact that most GA systems avoid only taking the fittest of the population in generating the next but rather a random (or semi-random) selection with a weighting toward those that are fitter.

For different genome types, different mutation types are suitable:

- *Bit string mutation:* The mutation of bit strings ensue through bit flips at random positions.

Example:

1	0	1	0	0	1	0
				↓		
1	0	1	0	1	1	0

The probability of a mutation of a bits is $\frac{1}{l}$, where l is the length of the binary vector. Thus, a mutation rate of 1 per mutation and individual selected for mutation is reached.

- *Flip Bit:* This mutation operator takes the chosen genome and invert the bits. (i.e. if the genome bit is 1,it is changed to 0 and vice versa)
- *Boundary:* This mutation operator replaces the genome with either lower or upper bound randomly. This can be used for integer and float genes.

- *Non-Uniform:* The probability that amount of mutation will go to 0 with the next generation is increased by using non-uniform mutation operator. It keeps the population from stagnating in the early stages of the evolution. It tunes solution in later stages of evolution. This mutation operator can only be used for integer and float genes.
- *Uniform:* This operator replaces the value of the chosen gene with a uniform random value selected between the user-specified upper and lower bounds for that gene. This mutation operator can only be used for integer and float genes.
- *Gaussian:* This operator adds a unit Gaussian distributed random value to the chosen gene. If it falls outside of the user-specified lower or upper bounds for that gene, the new gene value is clipped. This mutation operator can only be used for integer and float genes.

6

Parents Fields of Genetic Algorithms

Evolutionary Algorithm

In artificial intelligence, an evolutionary algorithm (EA) is a subset of evolutionary computation, a generic population-based metaheuristic optimisation algorithm. An EA uses some mechanisms inspired by biological evolution: reproduction, mutation, recombination, and selection. Candidate solutions to the optimisation problem play the role of individuals in a population, and the fitness function determines the environment within which the solutions "live". Evolution of the population then takes place after the repeated application of the above operators. Artificial evolution (AE) describes a process involving individual evolutionary algorithms; EAs are individual components that participate in an AE.

Evolutionary algorithms often perform well approximating solutions to all types of problems because they ideally do not make any assumption about the underlying fitness landscape; this generality is shown by successes in fields as diverse as engineering, art, biology, economics, marketing, genetics, operations research, robotics, social sciences, physics, politics and chemistry.

Apart from their use as mathematical optimisers, evolutionary computation and algorithms have also been used as an experimental framework within which to validate theories about biological evolution and natural selection, particularly through work in the field of artificial life. Techniques from evolutionary algorithms applied to the modelling of biological evolution are generally limited to explorations of microevolutionary processes, however some computer simulations, such as Tierra and Avida, attempt to model macroevolutionary dynamics. In most of real applications of EAs, computational complexity is a prohibiting factor.

In fact, this computational complexity is due to fitness function evaluation. Fitness approximation is one of the solutions to overcome this difficulty. However, seemingly simple EA can solve often complex problems; therefore, there may be no direct link between algorithm complexity and problem complexity.

Another possible limitation of many evolutionary algorithms is their lack of a clear genotype-phenotype distinction. In nature, the fertilized egg cell undergoes a complex process known as embryogenesis to become a mature phenotype. This indirect encoding is believed to make the genetic search more robust (i.e. reduce the probability of fatal mutations), and also may improve the evolvability of the organism. Such indirect (aka generative or developmental) encodings also enable evolution to exploit the regularity in the environment. Recent work in the field of artificial embryogeny, or artificial developmental systems, seeks to address these concerns.

Implementation of Biological Processes

Usually, an initial population of randomly generated candidate solutions comprise the first generation. The fitness function is applied to the candidate solutions and any subsequent offspring.

In selection, parents for the next generation are chosen with a bias towards higher fitness. The parents reproduce one or two offsprings (new candidates) by copying their genes, with two possible changes: crossover recombines the parental genes and mutation alters the genotype of an individual in a random way. These new candidates compete with old candidates for their place in the next generation (survival of the fittest). This process can be repeated until a candidate with sufficient quality (a solution) is found or a previously defined computational limit is reached.

Evolutionary Algorithm Techniques

Similar techniques differ in the implementation details and the nature of the particular applied problem.

- Genetic algorithm - This is the most popular type of EA. One seeks the solution of a problem in the form of strings of numbers (traditionally binary, although the best representations are usually those that reflect something about the problem being solved), by applying operators such as recombination and mutation (sometimes one, sometimes both). This type of EA is often used in optimisation problems;

- Genetic programming - Here the solutions are in the form of computer programs, and their fitness is determined by their ability to solve a computational problem.
- Evolutionary programming - Similar to genetic programming, but the structure of the program is fixed and its numerical parameters are allowed to evolve;
- Evolution strategy - Works with vectors of real numbers as representations of solutions, and typically uses self-adaptive mutation rates;
- Neuroevolution - Similar to genetic programming but the genomes represent artificial neural networks by describing structure and connection weights. The genome encoding can be direct or indirect.

Related Techniques

Swarm algorithms, including:

- Ant colony optimisation - Based on the ideas of ant foraging by pheromone communication to form paths. Primarily suited for combinatorial optimisation and graph problems.
- Bees algorithm is based on the foraging behaviour of honey bees. It has been applied in many applications such as routing and scheduling.
- Cuckoo search is inspired by the brooding parasitism of some cuckoo species. It also uses Lévy flights, and thus it suits for global optimisation problems.
- Particle swarm optimisation - Based on the ideas of animal flocking behaviour. Also primarily suited for numerical optimisation problems.

Other population-based metaheuristic methods:

- Differential evolution - Based on vector differences and is therefore primarily suited for numerical optimisation problems.
- Firefly algorithm is inspired by the behaviour of fireflies, attracting each other by flashing light. This is especially useful for multimodal optimisation.
- Invasive weed optimisation algorithm - Based on the ideas of weed colony behaviour in searching and finding a suitable place for growth and reproduction.
- Harmony search - Based on the ideas of musicians' behaviour in searching for better harmonies. This algorithm is suitable for combinatorial optimisation as well as parameter optimisation.
- Gaussian adaptation - Based on information theory. Used for maximization of manufacturing yield, mean fitness or average information.

Evolutionary Computation

In computer science, evolutionary computation is a subfield of artificial intelligence (more particularly computational intelligence) that involves combinatorial optimisation problems.

Evolutionary computation uses iterative progress, such as growth or development in a population. This population is then selected in a guided random search using parallel processing to achieve the desired end. Such processes are often inspired by biological mechanisms of evolution.

As evolution can produce highly optimised processes and networks, it has many applications in computer science. Here, simulations of evolution using evolutionary algorithms and artificial life started with the work of Nils Aall Barricelli in the 1960s, and was extended by Alex Fraser, who published a series of papers on simulation of artificial selection. Artificial evolution became a widely recognised optimisation method as a result of the work of Ingo Rechenberg in the 1960s and early 1970s, who used evolution strategies to solve complex engineering problems.

Genetic algorithms in particular became popular through the writing of John Holland. As academic interest grew, dramatic increases in the power of computers allowed practical applications, including the automatic evolution of computer programs. Evolutionary algorithms are now used to solve multidimensional problems more efficiently than software produced by human designers, and also to optimise the design of systems.

History

The use of Darwinian principles for automated problem solving originated in the fifties. It was not until the sixties that three distinct interpretations of this idea started to be developed in three different places. Evolutionary programming was introduced by Lawrence J. Fogel in the USA, while John Henry Holland called his method a genetic algorithm. In Germany Ingo Rechenberg and Hans-Paul Schwefel introduced evolution strategies.

These areas developed separately for about 15 years. From the early nineties on they are unified as different representatives ("dialects") of one technology, called evolutionary computing. Also in the early nineties, a fourth stream following the general ideas had emerged – genetic programming. Since the 1990s, evolutionary computation has largely become swarm-based computation, and nature-inspired algorithms are becoming an increasingly significant part.

These terminologies denote the field of evolutionary computing and consider evolutionary programming, evolution strategies, genetic algorithms, and genetic programming as sub-areas.

Techniques

Evolutionary computing techniques mostly involve metaheuristic optimisation algorithms. Broadly speaking, the field includes:

Evolutionary algorithms:

- Genetic algorithms
- Evolutionary programming
- Evolution strategy
- Eagle strategy
- Swarm intelligence
- Ant colony optimisation
- Particle swarm optimisation

and in a lesser extent also:

- Artificial life
- Artificial immune systems
- Cuckoo search
- Cultural algorithms
- Differential evolution
- Firefly algorithm
- Harmony search
- Learning classifier systems
- Learnable Evolution Model
- Parallel simulated annealing
- Self-organization such as self-organizing maps, competitive learning
- Self-Organizing Migrating Genetic Algorithm
- Swarm-based computing

Evolutionary Algorithms

Evolutionary algorithms form a subset of evolutionary computation in that they generally only involve techniques implementing mechanisms inspired by biological evolution such as reproduction, mutation, recombination, natural selection and survival of the fittest.

Candidate solutions to the optimisation problem play the role of individuals in a population, and the cost function determines the environment within which the solutions "live". Evolution of the population then takes place after the repeated application of the above operators.

In this process, there are two main forces that form the basis of evolutionary systems: Recombination and mutation create the necessary diversity and thereby facilitate novelty, while selection acts as a force increasing quality.

Many aspects of such an evolutionary process are stochastic. Changed pieces of information due to recombination and mutation are randomly chosen. On the other hand, selection operators can be either deterministic, or stochastic. In the latter case, individuals with a higher fitness have a higher chance to be selected than individuals with a lower fitness, but typically even the weak individuals have a chance to become a parent or to survive.

Metaheuristic

In computer science, metaheuristic designates a computational method that optimises a problem by iteratively trying to improve a candidate solution with regard to a given measure of quality. Metaheuristics make few or no assumptions about the problem being optimised and can search very large spaces of candidate solutions.

However, metaheuristics do not guarantee an optimal solution is ever found. Many metaheuristics implement some form of stochastic optimisation. Other terms having a similar meaning as metaheuristic, are: derivative-free, direct search, black-box, or indeed just heuristic optimiser. Several books and survey papers have been published on the subject.

Applications

Metaheuristics are used for combinatorial optimisation in which an optimal solution is sought over a discrete search-space. An example problem is the travelling salesman problem where the search-space of candidate solutions grows more than exponentially as the size of the problem increases, which makes an exhaustive search for the optimal solution infeasible. Additionally, multidimensional combinatorial problems, including most design problems in engineering such as form-finding and behaviour-finding, suffer from the curse of dimensionality, which also makes them infeasible for exhaustive search or analytical methods. Popular metaheuristics for combinatorial problems include simulated annealing by Kirkpatrick et al., genetic algorithms by Holland et al., ant colony optimisation by Dorigo, scatter search and tabu search by Glover.

Metaheuristics are also used for problems over real-valued search-spaces, where the classic way of optimisation is to derive the gradient of the function to be optimised and then employ gradient descentor a quasi-Newton method. Metaheuristics do not use the gradient or Hessian matrix so their advantage is that the function to be optimised need not be continuous or differentiable and it can also have constraints. Popular metaheuristic optimisers for real-valued search-spaces include particle swarm optimisation by Eberhart and Kennedy, differential evolution by Storn and Price and evolution strategies by Rechenberg and Schwefel.

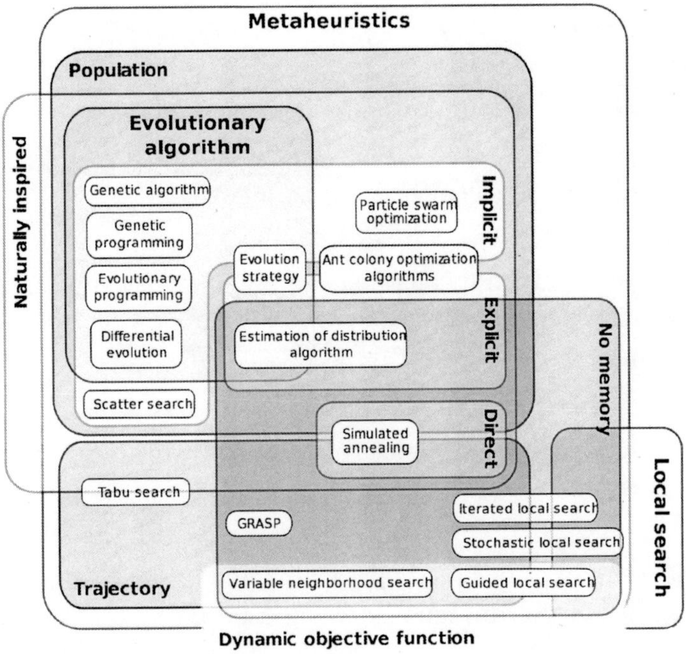

Figure: *Different classifications of metaheuristics.*

Metaheuristics based on decomposition techniques have also been proposed for tackling hard combinatorial problems of large size.

Main Contributions

Many different metaheuristics are in existence and new variants are continually being proposed. Some of the most significant contributions to the field are:

- 1952: Robbins and Monro work on stochastic optimisation methods.
- 1952: Fermi and Metropolis develop an early form of pattern search as described belatedly by Davidon.
- 1954: Barricelli carry out the first simulations of the evolution process and use them on general optimisation problems.
- 1963: Rastrigin proposes random search.
- 1965: Matyas proposes random optimisation.
- 1965: Rechenberg proposes evolution strategies.

- 1965: Nelder and Mead propose a simplex heuristic, which was shown by Powell to converge to non-stationary points on some problems.
- 1966: Fogel et al. propose evolutionary programming.
- 1970: Hastings proposes the Metropolis-Hastings algorithm.
- 1970: Cavicchio proposes adaptation of control parameters for an optimiser.
- 1970: Kernighan and Lin propose a graph partitioning method, related to variable-depth search and prohibition-based (tabu) search.
- 1975: Holland proposes the genetic algorithm.
- 1977: Glover proposes Scatter Search.
- 1978: Mercer and Sampson propose a metaplan for tuning an optimiser's parameters by using another optimiser.
- 1980: Smith describes genetic programming.
- 1983: Kirkpatrick et al. propose simulated annealing.
- 1986: Glover proposes tabu search, first mention of the term metaheuristic.
- 1986: Farmer et al. work on the artificial immune system.
- 1986: Grefenstette proposes another early form of metaplan for tuning an optimiser's parameters by using another optimiser.
- 1988: First conference on genetic algorithms is organized at the University of Illinois at Urbana-Champaign.
- 1988: Koza registers his first patent on genetic programming.
- 1989: Goldberg publishes a well known book on genetic algorithms.
- 1989: Evolver, the first optimisation software using the genetic algorithm.
- 1989: Moscato proposes the memetic algorithm.
- 1991: Interactive evolutionary computation.
- 1992: Dorigo proposes the ant colony algorithm.
- 1993: The journal, Evolutionary Computation, begins publication by the Massachusetts Institute of Technology.
- 1993: Fonseca and Fleming propose MOGA for multiobjective optimisation.
- 1994: Battiti and Tecchiolli propose Reactive Search Optimisation(RSO) principles for the online self-tuning of heuristics.
- 1994: Srinivas and Deb propose NSGA for multiobjective optimisation.
- 1995: Kennedy and Eberhart propose particle swarm optimisation.

- 1995: Wolpert and Macready prove the no free lunch theorems.
- 1996: Mühlenbein and Paaß work on the estimation of distribution algorithm.
- 1996: Hansen and Ostermeier propose CMA-ES.
- 1997: Storn and Price propose differential evolution.
- 1997: Rubinstein proposes the cross entropy method.
- 1999: Taillard and Voss propose POPMUSIC.
- 2001: Geem et al. propose harmony search.
- 2002: Deb et al. propose NSGA-II for multiobjective optimisation.
- 2004: Nakrani and Tovey propose bee colony optimisation.
- 2005: Krishnanand and Ghose propose Glowworm swarm optimisation.
- 2005: Karaboga proposes Artificial Bee Colony Algorithm (ABC).
- 2006: Haddad et al. introduces honey-bee mating optimisation.
- 2007: Hamed Shah-Hosseini introduces Intelligent Water Drops.
- 2008: Wierstra et al. propose natural evolution strategies based on the natural gradient.
- 2008: Yang introduces firefly algorithm.
- 2008: Mucherino and Seref propose the Monkey Search
- 2009: Yang and Deb introduce cuckoo search.
- 2011: Hamed Shah-Hosseini proposes the Galaxy-based Search Algorithm.
- 2011: Tamura and Yasuda propose spiral optimisation.

Criticism

Mathematical analyses of metaheuristics have been presented in the literature, the work by Trelea, amongst others, for analysis of particle swarm optimisation, and Zaharie for analysis of differential evolution. These analyses make a number of assumptions in regard to the optimiser variants and the simplicity of the optimisation problems which limit their validity in real-world optimisation scenarios. Performance and convergence aspects of metaheuristic optimisers are therefore often demonstrated empirically in the research literature. This has been criticized in the no free lunch set of theorems by Wolpert and Macready, which, among other things, prove that all optimisers perform equally well when averaged over all problems. The practical relevance of the no free lunch theorems however is minor, because they will generally not hold on the collection of problems a practitioner is facing.

For the practitioner the most relevant issue is that metaheuristics are not guaranteed to find the optimum or even a satisfactory near-optimal solution. All metaheuristics will eventually encounter problems on which they perform poorly and the practitioner must gain experience in which optimisers work well on different classes of problems.

Stochastic Optimisation

Stochastic optimisation (SO) methods are optimisation methods that generate and use random variables. For stochastic problems, the random variables appear in the formulation of the optimisation problem itself, which involve random objective functions or random constraints, for example.

Stochastic optimisation methods also include methods with random iterates. Some stochastic optimisation methods use random iterates to solve stochastic problems, combining both meanings of stochastic optimisation. Stochastic optimisation methods generalise deterministic methods for deterministic problems.

Methods for Stochastic Functions

Partly random input data arise in such areas as real-time estimation and control, simulation-based optimisation where Monte Carlo simulations are run as estimates of an actual system, and problems where there is experimental (random) error in the measurements of the criterion.

In such cases, knowledge that the function values are contaminated by random "noise" leads naturally to algorithms that use statistical inference tools to estimate the "true" values of the function and/or make statistically optimal decisions about the next steps. Methods of this class include

- stochastic approximation (SA), by Robbins and Monro (1951)
- stochastic gradient descent
- finite-difference SA by Kiefer and Wolfowitz (1952)
- simultaneous perturbation SA by Spall (1992)

Randomized Search Methods

On the other hand, even when the data set consists of precise measurements, some methods introduce randomness into the search-process to accelerate progress. Such randomness can also make the method less sensitive to modelling errors. Further, the injected randomness may enable the method to escape a local minimum and eventually to approach a global optimum.

Indeed, this randomization principle is known to be a simple and effective way to obtain algorithms with almost certain good performance uniformly across many data sets, for many sorts of problems. Stochastic optimisation methods of this kind include:

- simulated annealing by S. Kirkpatrick, C. D. Gelatt and M. P. Vecchi (1983)
- reactive search optimisation (RSO) by Roberto Battiti, G. Tecchiolli (1994), recently reviewed in the reference book
- cross-entropy method by Rubinstein and Kroese (2004)
- random search by Anatoly Zhigljavsky (1991)
- stochastic tunnelling
- parallel tempering a.k.a. replica exchange
- stochastic hill climbing
- swarm algorithms
- evolutionary algorithms
- genetic algorithms by Holland (1975)
- evolution strategies

Mathematical Optimisation

In mathematics, computational science, or management science, mathematical optimisation (alternatively, optimisation or mathematical programming) refers to the selection of a best element from some set of available alternatives.

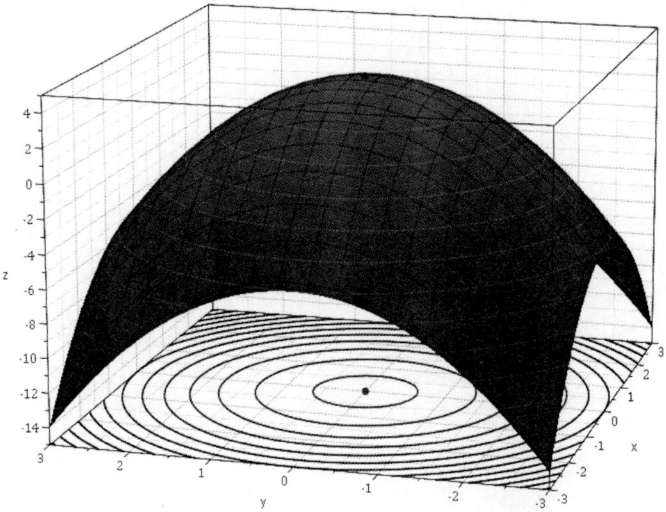

Figure: *The maximum of a paraboloid (red dot)*

In the simplest case, an optimisation problem consists of maximizing or minimizing a real function by systematically choosing input values from within an allowed set and computing the value of the function. The generalisation of optimisation theory and techniques to other formulations comprises a large area of applied mathematics. More generally, optimisation includes finding "best available" values of some objective function given a defined domain, including a variety of different types of objective functions and different types of domains.

Optimisation Problems

An optimisation problem can be represented in the following way

Given: a function f : A R from some set A to the real numbers

Sought: an element x_0 in A such that $f(x_0) \leq f(x)$ for all x in A ("minimization") or such that $f(x_0) \geq f(x)$ for all x in A ("maximization").

Such a formulation is called an optimisation problem or a mathematical programming problem (a term not directly related to computer programming, but still in use for example in linear programming). Many real-world and theoretical problems may be modelled in this general framework. Problems formulated using this technique in the fields of physics and computer vision may refer to the technique as energy minimization, speaking of the value of the function f as representing the energy of the system being modelled.

Typically, A is some subset of the Euclidean space R, often specified by a set of constraints, equalities or inequalities that the members of A have to satisfy. The domain A of f is called the search space or the choice set, while the elements of A are called candidate solutions or feasible solutions.

The function f is called, variously, an objective function, cost function (minimization), utility function (maximization), or, in certain fields, energy function, or energy functional. A feasible solution that minimizes (or maximizes, if that is the goal) the objective function is called an optimal solution.

By convention, the standard form of an optimisation problem is stated in terms of minimization. Generally, unless both the objective function and the feasible region are convex in a minimization problem, there may be several local minima, where a local minimum x is defined as a point for which there exists some $\delta > 0$ so that for all x such that

$$\left\| x - x^* \right\| \leq \delta;$$

the expression

$$f\left(x^*\right) \leq f\left(x\right)$$

holds; that is to say, on some region around x all of the function values are greater than or equal to the value at that point. Local maxima are defined similarly.

A large number of algorithms proposed for solving non-convex problems – including the majority of commercially available solvers – are not capable of making a distinction between local optimal solutions and rigorous optimal solutions, and will treat the former as actual solutions to the original problem. The branch of applied mathematics and numerical analysis that is concerned with the development of deterministic algorithms that are capable of guaranteeing convergence in finite time to the actual optimal solution of a non-convex problem is called global optimisation.

Notation

Optimisation problems are often expressed with special notation. Here are some examples.

Minimum and Maximum Value of a Function

Consider the following notation:

$$\min_{x\in\mathbb{R}}\left(x^2+1\right)$$

This denotes the minimum value of the objective function x + 1, when choosing x from the set of real numbers $\mathbb{R}$. The minimum value in this case is 1, occurring at x = 0.

Similarly, the notation

$$\max_{x\in\mathbb{R}} 2x$$

asks for the maximum value of the objective function 2x, where x may be any real number. In this case, there is no such maximum as the objective function is unbounded, so the answer is “infinity” or “undefined”.

Optimal Input Arguments

Consider the following notation:

$$\underset{x\in(-\infty,-1)}{\arg\min}\ x^2+1,$$

or equivalently

$$\underset{x}{\arg\min}\, x^2+1, \textit{subject to}: x\in\left(-\infty,-1\right).$$

This represents the value (or values) of the argument x in the interval $(-\infty,-1]$ that minimizes (or minimize) the objective function x + 1 (the actual minimum value of that function is not what the problem asks for). In this case, the answer is x = –1, since x = 0 is infeasible, i.e. does not belong to the feasible set.

Similarly,

$$\underset{x\varepsilon[-5,5],\, y\in\mathbb{R}}{\arg\max} \quad x\cos(y)$$

or equivalently

$$\underset{x,y}{\arg\max}\ x\cos(y), \textit{subject to}: x\in[-5,5], \quad y\in\mathbb{R}$$

represents the (x,y) pair (or pairs) that maximizes (or maximize) the value of the objective function xcos(y), with the added constraint that x lie in the interval [– 5,5] (again, the actual maximum value of the expression does not matter). In this case, the solutions are the pairs of the form (5, 2kπ) and (–5,(2k+1)π), where k ranges over all integers.

Arg min and arg max are sometimes also written argmin and argmax, and stand for argument of the minimum and argument of the maximum.

History

Fermat and Lagrange found calculus-based formulas for identifying optima, while Newton and Gaussproposed iterative methods for moving towards an optimum. Historically, the first term for optimisation was "linear programming", which was due to George B. Dantzig, although much of the theory had been introduced by Leonid Kantorovich in 1939. Dantzig published the Simplex algorithm in 1947, and John von Neumann developed the theory of duality in the same year.

The term programming in this context does not refer to computer programming. Rather, the term comes from the use of program by the United States military to refer to proposed training and logistics schedules, which were the problems Dantzig studied at that time. Later important researchers in mathematical optimisation include the following:

- Richard Bellman
- Ronald A. Howard
- Narendra Karmarkar
- William Karush
- Leonid Khachiyan
- Bernard Koopman
- Harold Kuhn
- Joseph Louis Lagrange

- László Lovász
- Arkadii Nemirovskii
- Yurii Nesterov
- Boris Polyak
- Lev Pontryagin
- James Renegar
- R. Tyrrell Rockafellar
- Cornelis Roos
- Naum Z. Shor
- Michael J. Todd
- Albert Tucker.

Major Subfields

- Convex programming studies the case when the objective function is convex (minimization) or concave (maximization) and the constraint set is convex. This can be viewed as a particular case of nonlinear programming or as generalisation of linear or convex quadratic programming.
- Linear programming (LP), a type of convex programming, studies the case in which the objective function f is linear and the set of constraints is specified using only linear equalities and inequalities. Such a set is called a polyhedron or a polytope if it is bounded.
- Second order cone programming (SOCP) is a convex program, and includes certain types of quadratic programs.
- Semidefinite programming (SDP) is a subfield of convex optimisation where the underlying variables are semidefinite matrices. It is generalisation of linear and convex quadratic programming.
- Conic programming is a general form of convex programming. LP, SOCP and SDP can all be viewed as conic programs with the appropriate type of cone.
- Geometric programming is a technique whereby objective and inequality constraints expressed as posynomials and equality constraints as monomials can be transformed into a convex program.
- Integer programming studies linear programs in which some or all variables are constrained to take on integer values. This is not convex, and in general much more difficult than regular linear programming.

- Quadratic programming allows the objective function to have quadratic terms, while the feasible set must be specified with linear equalities and inequalities. For specific forms of the quadratic term, this is a type of convex programming.
- Fractional programming studies optimisation of ratios of two nonlinear functions. The special class of concave fractional programs can be transformed to a convex optimisation problem.
- Nonlinear programming studies the general case in which the objective function or the constraints or both contain nonlinear parts. This may or may not be a convex program. In general, whether the program is convex affects the difficulty of solving it.
- Stochastic programming studies the case in which some of the constraints or parameters depend on random variables.
- Robust programming is, like stochastic programming, an attempt to capture uncertainty in the data underlying the optimisation problem. This is not done through the use of random variables, but instead, the problem is solved taking into account inaccuracies in the input data.
- Combinatorial optimisation is concerned with problems where the set of feasible solutions is discrete or can be reduced to a discrete one.
- Infinite-dimensional optimisation studies the case when the set of feasible solutions is a subset of an infinite-dimensional space, such as a space of functions.
- Metaheuristics make few or no assumptions about the problem being optimised and can search very large spaces of candidate solutions. However, metaheuristics do not guarantee an optimal solution is ever found.
- Constraint satisfaction studies the case in which the objective function f is constant (this is used in artificial intelligence, particularly in automated reasoning).
- Constraint programming.
- Disjunctive programming is used where at least one constraint must be satisfied but not all. It is of particular use in scheduling.

In a number of subfields, the techniques are designed primarily for optimisation in dynamic contexts (that is, decision making over time):

- Calculus of variations seeks to optimise an objective defined over many points in time, by considering how the objective function changes if there is a small change in the choice path.

- Optimal control theory is a generalisation of the calculus of variations.
- Dynamic programming studies the case in which the optimisation strategy is based on splitting the problem into smaller subproblems. The equation that describes the relationship between these subproblems is called the Bellman equation.
- Mathematical programming with equilibrium constraints is where the constraints include variational inequalities or complementarities.

Multi-objective Optimisation

Adding more than one objective to an optimisation problem adds complexity. For example, to optimise a structural design, one would want a design that is both light and rigid. Because these two objectives conflict, a trade-off exists. There will be one lightest design, one stiffest design, and an infinite number of designs that are some compromise of weight and stiffness. The set of trade-off designs that cannot be improved upon according to one criterion without hurting another criterion is known as the Pareto set. The curve created plotting weight against stiffness of the best designs is known as the Pareto frontier.

A design is judged to be "Pareto optimal" (equivalently, "Pareto efficient" or in the Pareto set) if it is not dominated by any other design: If it is worse than another design in some respects and no better in any respect, then it is dominated and is not Pareto optimal.

Multi-modal Optimisation

Optimisation problems are often multi-modal; that is they possess multiple good solutions. They could all be globally good (same cost function value) or there could be a mix of globally good and locally good solutions. Obtaining all (or at least some of) the multiple solutions is the goal of a multi-modal optimiser. Classical optimisation techniques due to their iterative approach do not perform satisfactorily when they are used to obtain multiple solutions, since it is not guaranteed that different solutions will be obtained even with different starting points in multiple runs of the algorithm. Evolutionary Algorithmsare however a very popular approach to obtain multiple solutions in a multi-modal optimisation task.

Classification of Critical Points and Extrema

Feasibility Problem

The satisfiability problem, also called the feasibility problem, is just the problem of finding any feasible solution at all without regard to objective value. This can be regarded as the special case of mathematical optimisation where the objective value is the same for every solution, and thus any solution is optimal. Many optimisation algorithms need to start from a feasible point. One way to obtain such a point is to relax the feasibility conditions using a slack variable; with enough slack, any starting point is feasible. Then, minimize that slack variable until slack is null or negative.

Existence

The extreme value theorem of Karl Weierstrass states that a continuous real-valued function on a compact set attains its maximum and minimum value. More generally, a lower semi-continuous function on a compact set attains its minimum; an upper semi-continuous function on a compact set attains its maximum.

Necessary Conditions for Optimality

One of Fermat's theorems states that optima of unconstrained problems are found at stationary points, where the first derivative or the gradient of the objective function is zero.

More generally, they may be found at critical points, where the first derivative or gradient of the objective function is zero or is undefined, or on the boundary of the choice set. An equation (or set of equations) stating that the first derivative(s) equal(s) zero at an interior optimum is called a 'first-order condition' or a set of first-order conditions.

Optima of inequality-constrained problems are instead found by the Lagrange multiplier method. This method calculates a system of inequalities called the 'Karush–Kuhn–Tucker conditions' or 'complementary slackness conditions', which may then be used to calculate the optimum.

Sufficient Conditions for Optimality

While the first derivative test identifies points that might be optima, this test does not distinguish a point which is a minimum from one that is a maximum or one that is neither.

When the objective function is twice differentiable, these cases can be distinguished by checking the second derivative or the matrix of second derivatives (called the Hessian matrix) in unconstrained problems, or the matrix of second derivatives of the objective function and the constraints called the bordered Hessian in constrained problems.

The conditions that distinguish maxima, or minima, from other stationary points are called 'second-order conditions'. If a candidate solution satisfies the first-order conditions, then satisfaction of the second-order conditions as well is sufficient to establish at least local optimality.

Sensitivity and Continuity of Optima

The envelope theorem describes how the value of an optimal solution changes when an underlying parameter changes. The process of computing this change is called comparative statics.

The maximum theorem of Claude Berge (1963) describes the continuity of an optimal solution as a function of underlying parameters.

Calculus of Optimisation

For unconstrained problems with twice-differentiable functions, some critical points can be found by finding the points where the gradient of the objective function is zero (that is, the stationary points).

More generally, a zero subgradient certifies that a local minimum has been found for minimization problems with convex functions and other locally Lipschitz functions.

Further, critical points can be classified using the definiteness of the Hessian matrix: If the Hessian is positive definite at a critical point, then the point is a local minimum; if the Hessian matrix is negative definite, then the point is a local maximum; finally, if indefinite, then the point is some kind of saddle point.

Constrained problems can often be transformed into unconstrained problems with the help of Lagrange multipliers. Lagrangian relaxation can also provide approximate solutions to difficult constrained problems.

When the objective function is convex, then any local minimum will also be a global minimum. There exist efficient numerical techniques for minimizing convex functions, such as interior-point methods.

Computational Optimisation Techniques

To solve problems, researchers may use algorithms that terminate in a finite number of steps, oriterative methods that converge to a solution (on some specified class of problems), or heuristics that may provide approximate solutions to some problems (although their iterates need not converge).

Optimisation Algorithms

- Simplex algorithm of George Dantzig, designed for linear programming.
- Extensions of the simplex algorithm, designed for quadratic programming and for linear-fractional programming.
- Variants of the simplex algorithm that are especially suited for network optimisation.
- Combinatorial algorithms

Iterative Methods

The iterative methods used to solve problems of nonlinear programming differ according to whether they evaluate Hessians, gradients, or only function values. While evaluating Hessians (H) and gradients (G) improves the rate of convergence, such evaluations increase the computational complexity (or computational cost) of each iteration. In some cases, the computational complexity may be excessively high.

One major criterion for optimisers is just the number of required function evaluations as this often is already a large computational effort, usually much more effort than within the optimiser itself, which mainly has to operate over the N variables.

The derivatives provide detailed information for such optimisers, but are even harder to calculate, e.g. approximating the gradient takes at least N+1 function evaluations. For approximations of the 2nd derivatives (collected in the Hessian matrix) the number of function evaluations is in the order of N^2. Newton's method requires the 2nd order derivates, so for each iteration the number of function calls is in the order of N^2, but for a simpler pure gradient optimiser it is only N.

However, gradient optimisers need usually more iterations than Newton's algorithm. Which one is best wrt. number of function calls depends on the problem itself.

- Methods that evaluate Hessians (or approximate Hessians, using finite differences):
- Newton's method
- Sequential quadratic programming: A method for small-medium scale constrained problems. Some versions can handle large-dimensional problems.
- Methods that evaluate gradients or approximate gradients using finite differences (or even subgradients):
- Quasi-Newton methods: Iterative methods for medium-large problems (e.g. $N<1000$).
- Conjugate gradient methods: Iterative methods for large problems. (In theory, these methods terminate in a finite number of steps with quadratic objective functions, but this finite termination is not observed in practice on finite–precision computers.)
- Interior point methods: This is a large class of methods for constrained optimisation. Some interior-point methods use only (sub)gradient information, and others of which require the evaluation of Hessians.

- Gradient descent (alternatively, "steepest descent" or "steepest ascent"): A (slow) method of historical and theoretical interest, which has had renewed interest for finding approximate solutions of enormous problems.
- Subgradient methods - An iterative method for large locally Lipschitz functions using generalised gradients. Following Boris T. Polyak, subgradient–projection methods are similar to conjugate–gradient methods.
- Bundle method of descent: An iterative method for small–medium sized problems with locally Lipschitz functions, particularly for convex minimization problems. (Similar to conjugate gradient methods)
- Ellipsoid method: An iterative method for small problems with quasiconvex objective functions and of great theoretical interest, particularly in establishing the polynomial time complexity of some combinatorial optimisation problems. It has similarities with Quasi-Newton methods.
- Reduced gradient method (Frank–Wolfe) for approximate minimization of specially structured problems with linear constraints, especially with traffic networks. For general unconstrained problems, this method reduces to the gradient method, which is regarded as obsolete (for almost all problems).
- Methods that evaluate only function values: If a problem is continuously differentiable, then gradients can be approximated using finite differences, in which case a gradient-based method can be used. Usually such non-gradient methods work only fine for problems with few parameter only.
- Interpolation methods
- Pattern search methods, which have better convergence properties than the Nelder–Mead simplicial heuristic (listed below, under heuristics).

Global Convergence

More generally, if the objective function is not a quadratic function, then many optimisation methods use other methods to ensure that some subsequence of iterations converges to an optimal solution. The first and still popular method for ensuring convergence relies on-line searches, which optimise a function along one dimension. A second and increasingly popular method for ensuring convergence uses trust regions. Both line searches and trust regions are used in modern methods of non-differentiable optimisation. Usually a global optimiser is much slower than advanced local optimisers (such as

BFGS), so often an efficient global optimiser can be constructed by starting the local optimiser from different starting points.

Heuristics

Besides (finitely terminating) algorithms and (convergent) iterative methods, there are heuristics that can provide approximate solutions to some optimisation problems:

- Memetic algorithm
- Differential evolution
- Dynamic relaxation
- Genetic algorithms
- Hill climbing
- Nelder-Mead simplicial heuristic: A popular heuristic for approximate minimization (without calling gradients)
- Particle swarm optimisation
- Simulated annealing
- Tabu search

Applications

Mechanics and Engineering

Problems in rigid body dynamics (in particular articulated rigid body dynamics) often require mathematical programming techniques, since you can view rigid body dynamics as attempting to solve an ordinary differential equation on a constraint manifold; the constraints are various nonlinear geometric constraints such as "these two points must always coincide", "this surface must not penetrate any other", or "this point must always lie somewhere on this curve". Also, the problem of computing contact forces can be done by solving a linear complementarity problem, which can also be viewed as a QP (quadratic programming) problem. Many design problems can also be expressed as optimisation programs. This application is called design optimisation. One subset is the engineering optimisation, and another recent and growing subset of this field is multidisciplinary design optimisation, which, while useful in many problems, has in particular been applied to aerospace engineering problems.

Economics

Economics is closely enough linked to optimisation of agents that an influential definition relatedly describes economics qua science as the "study of human behaviour as a relationship between ends and scarce means" with alternative uses. Modern optimisation theory includes traditional optimisation theory but also overlaps with game theory and the study of economic equilibria. The Journal of Economic Literature codes classify mathematical programming, optimisation techniques, and related topics under JEL:C61-C63.

In microeconomics, the utility maximization problem and its dual problem, the expenditure minimization problem, are economic optimisation problems. Insofar as they behave consistently, consumers are assumed to maximize their utility, while firms are usually assumed to maximize their profit. Also, agents are often modelled as being risk-averse, thereby preferring to avoid risk. Asset prices are also modelled using optimisation theory, though the underlying mathematics relies on optimising stochastic processes rather than on static optimisation. Trade theory also uses optimisation to explain trade patterns between nations.

Since the 1970s, economists have modelled dynamic decisions over time using control theory. For example, microeconomists use dynamic search models to study labour-market behaviour. A crucial distinction is between deterministic and stochastic models. Macroeconomists build dynamic stochastic general equilibrium (DSGE) models that describe the dynamics of the whole economy as the result of the interdependent optimising decisions of workers, consumers, investors, and governments.

Operations Research

Another field that uses optimisation techniques extensively is operations research. Operations research also uses stochastic modelling and simulation to support improved decision-making. Increasingly, operations research uses stochastic programming to model dynamic decisions that adapt to events; such problems can be solved with large-scale optimisation and stochastic.

7

Applications of Genetic Algorithms

Computational Creativity

Computational creativity (also known as artificial creativity, mechanical creativity or creative computation) is a multidisciplinary endeavour that is located at the intersection of the fields of artificial intelligence, cognitive psychology, philosophy, and the arts. The goal of computational creativity is to model, simulate or replicate creativity using a computer, to achieve one of several ends:

- to construct a program or computer capable of human-level creativity
- to better understand human creativity and to formulate an algorithmic perspective on creative behaviour in humans
- to design programs that can enhance human creativity without necessarily being creative themselves

The field of computational creativity concerns itself with theoretical and practical issues in the study of creativity. Theoretical work on the nature and proper definition of creativity is performed in parallel with practical work on the implementation of systems that exhibit creativity, with one strand of work informing the other.

Theoretical Issues

As measured by the amount of activity in the field (e.g., publications, conferences and workshops), computational creativity is a growing area of research. But the field is still hampered by a number of fundamental problems:

- Creativity is very difficult, perhaps even impossible, to define in objective terms.
- Creativity takes many forms in human activity, some eminent (meaning "recognized" or "ingenious", e.g., Einstein's creativity; sometimes referred to as "Creativity" with a capital C) and some mundane.
- Creativity can mean different things in different contexts: Is it a state of mind, a talent or ability, or a process? Does it describe a person, an activity or an end-product? Can collaborative work in which exceptional products emerge from simple interactions be considered creative?

These are problems that complicate the study of creativity in general, but certain problems attach themselves specifically to computational creativity:

- Can creativity be hard-wired? In existing systems to which creativity is attributed, is the creativity that of the system or that of the system's programmer or designer?
- How do we evaluate computational creativity? What counts as creativity in a computational system? Are natural language generation systems creative? Are machine translation systems creative? What distinguishes research in computational creativity from research in artificial intelligence generally?
- If eminent creativity is about rule-breaking or the disavowal of convention, how is it possible for an algorithmic system to be creative? In essence, this is a variant of the Ada Lovelace objection to machine intelligence, as recapitulated by modern theorists such as Teresa Amabile: If a machine can do only what it was programmed to do, how can its behaviour ever be called creative?

Defining Creativity in Computational Terms

Since no single perspective or definition seems to offer a complete picture of creativity, the AI researchers Newell, Shaw and Simon developed the combination of novelty and usefulness into the cornerstone of a multi-pronged view of creativity, one that uses the following four criteria to categorize a given answer or solution as creative:

1. The answer is novel and useful (either for the individual or for society)
2. The answer demands that we reject ideas we had previously accepted
3. The answer results from intense motivation and persistence
4. The answer comes from clarifying a problem that was originally vague

Notice how these criteria touch on many of the stereotypical themes that are typically associated with creativity: newness and value (1), transformation and revolution (2), passion and drive (3), vision and insight (4). These four criteria also combine elements of the producer-perspective and the product-perspective described earlier: criterion (1) characterizes the two most important qualities of a creative product, while criteria (2) – (4) characterize the attitude and actions of the producer of such a product. A given product may satisfy all or none of these criteria, but we should expect products that exhibit all four to be widely perceived as creative, while products that exhibit just some of these criteria will be judged with greater subjectivity and variation. Though no criterion is likely to be either necessary or sufficient, criterion (1) is perhaps the most common hallmark of creativity and thus serves to anchor the others. From a computational perspective, then, one can consider (1) to be a must-have feature, and (2) – (4) as desirable extras.

Newell and Simon are best known for their contribution to the search-in-a-state-space paradigm of AI, sometimes caricatured as Good Old Fashioned AI (GOFAI), and it is interesting to consider how the GOFAI paradigm can incorporate these criteria. From a search perspective, criterion (1) characterizes the goal or end-state of a computational search, criterion (4) characterizes the starting state from which the search is launched, criterion (3) characterizes the scale of the search, suggesting that many dead-ends are likely to be encountered, while criterion (2) suggests that well-worn pathways through the search space are best avoided if a creative end-state is to be reached.

Key Ideas

Some high-level and philosophical themes recur throughout the field of computational creativity.

P-creativity and H-creativity

Margaret Boden refers to creativity that is novel merely to the agent that produces it as "P-creativity" (or "psychological creativity"), and refers to creativity that is recognized as novel by society at large as "H-creativity" (or "historical creativity").

Exploratory and Transformational Creativity

Boden also distinguishes between the creativity that arises from an exploration within an established conceptual space, and the creativity that arises from a deliberate transformation or transcendence of this space. She labels the former as "exploratory creativity" and the latter as "transformational creativity", seeing the latter as a form of creativity far more radical, challenging, and rarer than the former. Following Newell and Simon's criteria, we can see that both forms of creativity should produce results that are appreciably novel and useful (criterion 1), but exploratory creativity is more likely to arise from a thorough and persistent search of a well-understood space (criterion 3) while transformational creativity should involve the rejection of some of the constraints that define this space (criterion 2) or some of the assumptions that define the problem itself (criterion 4). Boden's insights have guided work in computational creativity at a very general level, providing more an inspirational touchstone for development work than a technical framework of algorithmic substance. However, Boden's insights are the subject of formalization, most notably in the work by Geraint Wiggins.

Generation and Evaluation

The criterion that creative products should be novel and useful means that creative computational systems are typically structured into two phases, generation and evaluation. In the first phase, novel (to the system itself, thus P-Creative) constructs are generated; unoriginal constructs that are already known to the system are filtered at this stage.

This body of potentially creative constructs are then evaluated, to determine which are meaningful and useful and which are not. This two-phase structure conforms to the Geneplore model of Finke, Ward and Smith, which is a psychological model of creative generation based on empirical observation of human creativity.

Combinatorial Creativity

A great deal, perhaps all, of human creativity can be understood as a novel combination of pre-existing ideas or objects. Common strategies for combinatorial creativity include:

- placing a familiar object in an unfamiliar setting (e.g., Marcel Duchamp's Fountain) or an unfamiliar object in a familiar setting (e.g., a fish-out-of-water story such as The Beverly Hillbillies)
- Blending two superficially different objects or genres (e.g., a sci-fi story set in the Wild West, with robot cowboys, as in Westworld, or the reverse, as in Firefly; Japanese haiku poems, etc.)
- Comparing a familiar object to a superficially unrelated and semantically distant concept (e.g., "Makeup is the Western burka"; "A zoo is a gallery with living exhibits")
- Adding a new and unexpected feature to an existing concept (e.g., adding a scalpel to a Swiss Army knife; adding a camera to a mobile phone)
- Compressing two incongruous scenarios into the same narrative to get a joke (e.g., the Emo Philips joke "Women are always using me to advance their careers. Damned anthropologists!")

- Using an iconic image from one domain in a domain for an unrelated or incongruous idea or product (e.g., using the Marlboro Man image to sell cars, or to advertise the dangers of smoking-related impotence).

The combinatorial perspective allows us to model creativity as a search process through the space of possible combinations. The combinations can arise from composition or concatenation of different representations, or through a rule-based or stochastic transformation of initial and intermediate representations. Genetic algorithms and neural networks can be used to generate blended or crossover representations that capture a combination of different inputs.

Bisociation

Arthur Koestler proposes a very general model of creative combination in his 1964 book The Act of Creation, claiming that scientific discovery, art and humour are all linked by a common mechanism called "bisociation". Koestler lacked a formal, computational vocabulary for describing bisociation, which he defined as a reconciliation of two orthogonal matrices of thought (conceptual structures, mental spaces).

Conceptual Blending

Mark Turner and Gilles Fauconnier propose a model called Conceptual Integration Networks that elaborates upon the ideas of Koestler by synthesizing ideas from Cognitive Linguistic research into mental spaces and conceptual metaphors. Their basic model defines an integration network as four connected spaces:

- A first input space (contains one conceptual structure or mental space)
- A second input space (to be blended with the first input)
- A generic space of stock conventions and image-schemas that allow the input spaces to be understood from an integrated perspective
- A blend space in which a selected projection of elements from both input spaces are combined; inferences arising from this combination also reside here, sometimes leading to emergent structures that conflict with the inputs.

Fauconnier and Turner describe a collection of optimality principles that are claimed to guide the construction of a well-formed integration network. In essence, they see blending as a compression mechanism in which two or more input structures are compressed into a single blend structure. This compression operates on the level of conceptual relations. For example, a series of similarity relations between the input spaces can be compressed into a single identity relationship in the blend.

Blending theory is an elaborate framework that provides a rich terminology for describing the products of creative thinking, from metaphors to jokes to neologisms to adverts. It is most typically applied retrospectively, to describe how a blended conceptual structure could have arisen from a particular pair of input structures. These conceptual structures are often good examples of human creativity, but blending theory is not a theory of creativity, nor – despite its authors' claims – does it describe a mechanism for creativity. The theory lacks an explanation for how a creative individual chooses the input spaces that should be blended to generate a desired result.

Nonetheless, some computational success has been achieved with the blending model by extending pre-existing computational models of analogical mapping that are compatible by virtue of their emphasis on connected semantic structures. More recently, Francisco Câmara Pereira presented an implementation of blending theory that employs ideas both from GOFAI and from genetic algorithms to realize some aspects of blending theory in a practical form; his example domains range from the linguistic to the visual, and the latter most notably includes the creation of mythical monsters by combining 3-D graphical models.

Linguistic Creativity

Language provides continuous opportunity for creativity, evident in the generation of novel sentences, phrasings, puns, neologisms, rhymes, allusions, sarcasm, irony, similes, metaphors, analogies, witticisms, and jokes. Native speakers of morphologically rich languages (including all Slavic languages) frequently create new word-forms that are easily understood, although they will never find their way to the dictionary. The area of natural language generation has been well studied, but these creative aspects of everyday language have yet to be incorporated with any robustness or scale.

Story Generation

Substantial work has been conducted in this area of linguistic creation since the 1970s, with the development of James Meehan's TALE-SPIN system. TALE-SPIN viewed stories as narrative descriptions of a problem-solving effort, and created stories by first establishing a goal for the story's characters so that their search for a solution could be tracked and recorded. The MINSTREL system represents a complex elaboration of this basis approach, distinguishing a range of character-level goals in the story from a range of author-level goals for the story. Systems like Bringsjord's BRUTUS elaborate these ideas further to create stories with complex inter-personal themes like betrayal. Nonetheless, MINSTREL explicitly models the creative process with a set of Transform Recall Adapt Methods (TRAMs) to create novel scenes from old. The MEXICA model of Rafael Pérez y Pérez and Mike Sharples is more explicitly interested in the creative process of storytelling, and implements a version of the engagement-reflection cognitive model of creative writing.

The company Narrative Science makes computer generated news and reports commercially available, including summarizing team sporting events based on statistical data from the game. It also creates financial reports and real estate analyses.

Metaphor and Simile

Example of a metaphor: "She was an ape."

Example of a simile: "Felt like a tiger-fur blanket." The computational study of these phenomena has mainly focused on interpretation as a knowledge-based process. Computationalists such as Yorick Wilks, James Martin, Dan Fass, John Barnden, and Mark Lee have developed knowledge-based approaches to the processing of metaphors, either at a linguistic level or a logical level. Tony Veale and Yanfen Hao have developed a system, called Sardonicus, that acquires a comprehensive database of explicit similes from the web; these similes are then tagged as bona-fide (e.g., "as hard as steel") or ironic (e.g., "as hairy as a bowling ball", "as pleasant as a root canal"); similes of either type can be retrieved on demand for any given adjective. They use these similes as the basis of an on-line metaphor generation system called Aristotle that can suggest lexical metaphors for a given descriptive goal (e.g., to describe a supermodel as skinny, the source terms "pencil", "whip", "whippet", "rope", "stick-insect" and "snake" are suggested).

Analogy

The process of analogical reasoning has been studied from both a mapping and a retrieval perspective, the latter being key to the generation of novel analogies. The dominant school of research, as advanced by Dedre Gentner. Views analogy as a structure-preserving process; this view has been implemented in the structure mapping engine or SME, the MAC/FAC retrieval engine (Many Are Called, Few Are Chosen), ACME (Analogical Constraint Mapping Engine) and ARCS (Analogical Retrieval Constraint System). Other mapping-based approaches include Sapper, which situates the mapping process in a semantic-network model of memory. Analogy is a very active sub-area of creative computation and creative cognition; active figures in this sub-area include Douglas Hofstadter, Paul Thagard, and Keith Holyoak. Also worthy of note here is Peter Turney and Michael Littman's machine learning approach to the solving of SAT-style analogy problems; their approach achieves a score that compares well with average scores achieved by humans on these tests.

Joke Generation

Humour is an especially knowledge-hungry process, and the most successful joke-generation systems to date have focussed on pun-generation, as exemplified by the work of Kim Binsted and Graeme Ritchie. This work includes the JAPE system, which can generate a wide range of puns that are consistently evaluated as novel and humorous by young children. An improved version of JAPE has been developed in the guise of the STANDUP system, which has been experimentally deployed as a means of enhancing linguistic interaction with children with communication disabilities. Some limited progress has been made in generating humour that involves other aspects of natural language, such as the deliberate misunderstanding of pronominal reference (in the work of Hans Wim Tinholt and Anton Nijholt), as well as in the generation of humorous acronyms in the HAHAcronym system of Oliviero Stock and Carlo Strapparava.

Neologisms

The blending of multiple word forms is a dominant force for new word creation in language; these new words are commonly called "blends" or "portmanteau words" (after Lewis Carroll). Tony Veale has developed a system called ZeitGeist that harvests neological headwords from Wikipedia and interprets them relative to their local context in Wikipedia and relative to specific word senses in WordNet. ZeitGeist has been extended to generate neologisms of its own; the approach combines elements from an inventory of word parts that are harvested from WordNet, and simultaneously determines likely glosses for these new words (e.g., "food traveller" for "gastronaut" and "time traveller" for "chrononaut"). It then uses Web search to determine which glosses are meaningful and which neologisms have not been used before; this search identifies the subset of generated words that are both novel ("H-creative") and useful. Neurolinguistic inspirations have been used to analyse the process of novel word creation in the brain, understand neurocognitive processes responsible for intuition, insight, imagination and creativity and to create a server that invents novel names for products, based on their description.

Poetry

Like jokes, poems involve a complex interaction of different constraints, and no general-purpose poem generator adequately combines the meaning, phrasing, structure and rhyme aspects of poetry. Nonetheless, Pablo Gervás has developed a noteworthy system called ASPERA that employs a case-based reasoning (CBR) approach to generating poetic formulations of a given input text via a composition of poetic fragments that are retrieved from a case-base of existing poems. Each poem fragment in the ASPERA case-base is annotated with a prose string that expresses the meaning of the fragment, and this prose string is used as the retrieval key for each fragment. Metrical rules are then used to combine these fragments into a well-formed poetic structure. Example software projects include:

- Racter
- Flowerewolf automatic poetry generator.

And poetry collections include:

- The Policeman's Beard Is Half Constructed.

Musical Creativity

Computational creativity in the music domain has focused both on the generation of musical scores for use by human musicians, and on the generation of music for performance by computers.

The domain of generation has included classical music (with software that generates music in the style of Mozartand Bach) and jazz. Most notably, David Cope has written a software system called "Experiments in Musical Intelligence" (or "EMI") that is capable of analysing and generalising from existing music by a human composer to generate novel musical compositions in the same style. EMI's output is convincing enough to persuade human listeners that its music is human-generated to a high level of competence.

Creativity research in jazz has focused on the process of improvisation and the cognitive demands that this places on a musical agent: reasoning about time, remembering and conceptualizing what has already been played, and planning ahead for what might be played next. The robot Shimon, developed by Gil Weinberg of Georgia Tech, has demonstrated jazz improvisation.

In the field of musical composition, the patented works by René-Louis Baron allowed to make a robot that can create and play a multitude of orchestrated melodies so-called "coherent" in any musical style. All outdoor physical parameter associated with one or more specific musical parameters, can influence and develop each of these songs. The patented invention Medal-Composerraises problems of copyright.

Visual and Artistic Creativity

Computational creativity in the generation of visual art has had some notable successes in the creation of both abstract art and representational art. The most famous program in this domain is Harold Cohen's AARON, which has been continuously developed and augmented since 1973. Though formulaic, Aaron exhibits a range of outputs, generating black-and-white drawings or colour paintings that incorporate human figures (such as dancers), potted plants, rocks, and other elements of background imagery. These images are of a sufficiently high quality to be displayed in reputable galleries. Other software artists of note include the NEvAr system (for "Neuro-Evolutionary Art") of Penousal Machado. NEvAr uses a genetic algorithm to derive a mathematical function that is then used to generate a coloured three-dimensional surface. A human user is allowed to select the best pictures after each phase of the genetic algorithm, and these preferences are used to guide successive phases, thereby pushing NEvAr's search into pockets of the search space that are considered most appealing to the user.

The Painting Fool, developed by Simon Colton originated as a system for overpainting digital images of a given scene in a choice of different painting styles, colour palettes and brush types. Given its dependence on an input source image to work with, the earliest iterations of the Painting Fool raised as questions about the extent of, or lack of, creativity in a computational art system. Nonetheless, in more recent work, The Painting Fool has been extended to create novel images, much as AARON does, from its own limited imagination. Images in this vein include cityscapes and forests, which are generated by a process of constraint satisfaction from some basic scenarios provided by the user (e.g., these scenarios allow the system to infer that objects closer to the viewing plane should be larger and more colour-saturated, while those further away should be less saturated and appear smaller). Artistically, the images now created by the Painting Fool appear on a par with those created by Aaron, though the extensible mechanisms employed by the former (constraint satisfaction, etc.) may well allow it to develop into a more elaborate and sophisticated painter.

Creativity in Problem Solving

Creativity is also useful in allowing for unusual solutions in problem solving. In psychology and cognitive science, this research area is called creative problem solving.

The Explicit-Implicit Interaction (EII) theory of creativity has recently been implemented using a CLARION-based computational model that allows for the simulation of incubation and insight in problem solving.

The emphasis of this computational creativity project is not on performance per se (as in artificial intelligence projects) but rather on the explanation of the psychological processes leading to human creativity and the reproduction of data collected in psychology experiments. So far, this project has been successful in providing an explanation for incubation effects in simple memory experiments, insight in problem solving, and reproducing the overshadowing effect in problem solving

Bioinformatics

Bioinformatics is the application of computer science and information technology to the field of biology and medicine. Bioinformatics deals with algorithms, databases and information systems, web technologies, artificial intelligence and soft computing, information and computation theory, software engineering, data mining, image processing, modelling and simulation, signal processing, discrete mathematics, control and system theory, circuit theory, and statistics, for generating new knowledge of biology and medicine, and improving & discovering new models of computation (e.g. DNA computing, neural computing, evolutionary computing, immuno-computing, swarm-computing, cellular-computing).

Commonly used software tools and technologies in this field include Java, XML, Perl, C, C++, Python, R, SQL,CUDA and MATLAB.

Bioinformatics was applied in the creation and maintenance of a database to store biological information at the beginning of the "genomic revolution", such as nucleotide sequences and amino acid sequences. Development of this type of database involved not only design issues but the development of complex interfaces whereby researchers could both access existing data as well as submit new or revised data.

In order to study how normal cellular activities are altered in different disease states, the biological data must be combined to form a comprehensive picture of these activities. Therefore, the field of bioinformatics has evolved such that the most pressing task now involves the analysis and interpretation of various types of data, including nucleotide and amino acid sequences, protein domains, and protein structures. The actual process of analysing and interpreting data is referred to as computational biology. Important sub-disciplines within bioinformatics and computational biology include:

- the development and implementation of tools that enable efficient access to, and use and management of, various types of information.
- the development of new algorithms (mathematical formulas) and statistics with which to assess relationships among members of large data sets, such as methods to locate a gene within a sequence, predict protein structure and/or function, and cluster protein sequences into families of related sequences.

The primary goal of bioinformatics is to increase the understanding of biological processes. What sets it apart from other approaches, however, is its focus on developing and applying computationally intensive techniques (e.g., pattern recognition, data mining, machine learning algorithms, and visualization) to achieve this goal. Major research efforts in the field include sequence alignment, gene finding, genome assembly, drug design, drug discovery, protein structure alignment, protein structure prediction, prediction of gene expression and protein–protein interactions, genome-wide association studies and the modelling of evolution.

Interestingly, the term bioinformatics was coined before the "genomic revolution". Paulien Hogewegand Ben Hesper introduced the term in 1978 to refer to "the study of information processes in biotic systems"., a definition that placed bioinformatics as a field parallel to biophysics or biochemistry (e.g. biochemistry is the study of chemical processes in biological systems) . However, its primary use since at least the late 1980s has been to describe the application of computer science and information sciences to the analysis of biological data, particularly in those areas of genomics involving large-scale DNA sequencing.

Bioinformatics now entails the creation and advancement of databases, algorithms, computational and statistical techniques and theory to solve formal and practical problems arising from the management and analysis of biological data. Over the past few decades rapid developments in genomic and other molecular research technologies and developments in information technologies have combined to produce a tremendous amount of information related to molecular biology. It is the name given to these mathematical and computing approaches used to glean understanding of biological processes.

Common activities in bioinformatics include mapping and analysing DNA and protein sequences, aligning different DNA and protein sequences to compare them and creating and viewing 3-D models of protein structures.

There are two fundamental ways of modelling a Biological system (e.g. living cell) both coming under Bioinformatic approaches.

- Static
- Sequences – Proteins, Nucleic acids and Peptides
- Structures – Proteins, Nucleic acids, Ligands (including metabolites and drugs) and Peptides
- Interaction data among the above entities including microarray data and Networks of proteins, metabolites
- Dynamic
- Systems Biology comes under this category including reaction fluxes and variable concentrations of metabolites
- Multi-Agent Based modelling approaches capturing cellular events such as signalling, transcription and reaction dynamics

A broad sub-category under bioinformatics is structural bioinformatics.

Major Research Areas

Sequence Analysis

Since the Phage Φ-X174 was sequenced in 1977, the DNA sequences of thousands of organisms have been decoded and stored in databases. This sequence information is analysed to determine genes that encode polypeptides (proteins), RNA genes, regulatory sequences, structural motifs, and repetitive sequences.

A comparison of genes within a species or between different species can show similarities between protein functions, or relations between species (the use of molecular systematics to construct phylogenetic trees). With the growing amount of data, it long ago became impractical to analyse DNA sequences manually.

Today, computer programs such as BLAST are used daily to search sequences from more than 260 000 organisms, containing over 190 billion nucleotides. These programs can compensate for mutations (exchanged, deleted or inserted bases) in the DNA sequence, to identify sequences that are related, but not identical. A variant of this sequence alignment is used in the sequencing process itself.

The so-called shotgun sequencing technique (which was used, for example, by The Institute for Genomic Research to sequence the first bacterial genome, Haemophilus influenzae) does not produce entire chromosomes, but instead generates the sequences of many thousands of small DNA fragments (ranging from 35 to 900 nucleotides long, depending on the sequencing technology).

The ends of these fragments overlap and, when aligned properly by a genome assembly program, can be used to reconstruct the complete genome. Shotgun sequencing yields sequence data quickly, but the task of assembling the fragments can be quite complicated for larger genomes.

For a genome as large as the human genome, it may take many days of CPU time on large-memory, multiprocessor computers to assemble the fragments, and the resulting assembly will usually contain numerous gaps that have to be filled in later. Shotgun sequencing is the method of choice for virtually all genomes sequenced today, and genome assembly algorithms are a critical area of bioinformatics research.

Another aspect of bioinformatics in sequence analysis is annotation, which involves computational gene finding to search for protein-coding genes, RNA genes, and other functional sequences within a genome. Not all of the nucleotides within a genome are part of genes. Within the genome of higher organisms, large parts of the DNA do not serve any obvious purpose. This so-called junk

DNA may, however, contain unrecognized functional elements. Bioinformatics helps to bridge the gap between genome and proteome projects — for example, in the use of DNA sequences for protein identification.

Genome Annotation

In the context of genomics, annotation is the process of marking the genes and other biological features in a DNA sequence. The first genome annotation software system was designed in 1995 by Dr. Owen White, who was part of the team at The Institute for Genomic Research that sequenced and analysed the first genome of a free-living organism to be decoded, the bacterium Haemophilus influenzae. Dr. White built a software system to find the genes (places in the DNA sequence that encode a protein), the transfer RNA, and other features, and to make initial assignments of function to those genes. Most current genome annotation systems work similarly, but the programs available for analysis of genomic DNA are constantly changing and improving.

Computational Evolutionary Biology

Evolutionary biology is the study of the origin and descent of species, as well as their change over time. Informatics has assisted evolutionary biologists in several key ways; it has enabled researchers to:

- trace the evolution of a large number of organisms by measuring changes in their DNA, rather than through physical taxonomy or physiological observations alone,
- more recently, compare entire genomes, which permits the study of more complex evolutionary events, such as gene duplication, horizontal gene transfer, and the prediction of factors important in bacterial speciation,
- build complex computational models of populations to predict the outcome of the system over time
- track and share information on an increasingly large number of species and organisms

Future work endeavours to reconstruct the now more complex tree of life. The area of research within computer science that uses genetic algorithms is sometimes confused with computational evolutionary biology, but the two areas are not necessarily related.

Literature Analysis

The growth in the number of published literature makes it virtually impossible to read every paper, resulting in disjointed subfields of research. Literature analysis aims to employ computational and statistical linguistics to mine this growing library of text resources. For example:

- abbreviation recognition - identify the long-form and abbreviation of biological terms,
- named entity recognition - recognizing biological terms such as gene names
- protein-protein interaction - identify which proteins interact with which proteins from text

The area of research draws from statistics and computational linguistics.

Analysis of Gene Expression

The expression of many genes can be determined by measuring mRNA levels with multiple techniques including microarrays, expressed cDNA sequence tag (EST) sequencing, serial analysis of gene expression (SAGE) tag sequencing, massively parallel signature sequencing (MPSS), RNA-Seq, also known as "Whole Transcriptome Shotgun Sequencing" (WTSS), or various applications of multiplexed in-situ hybridization.

All of these techniques are extremely noise-prone and/or subject to bias in the biological measurement, and a major research area in computational biology involves developing statistical tools to separate signal from noise in high-throughput gene expression studies. Such studies are often used to determine the genes implicated in a disorder: one might compare microarray data from cancerous epithelial cells to data from non-cancerous cells to determine the transcripts that are up-regulated and down-regulated in a particular population of cancer cells.

Analysis of Regulation

Regulation is the complex orchestration of events starting with an extracellular signal such as a hormone and leading to an increase or decrease in the activity of one or more proteins. Bioinformatics techniques have been applied to explore various steps in this process. For example, promoter analysis involves the identification and study of sequence motifs in the DNA surrounding the coding region of a gene.

These motifs influence the extent to which that region is transcribed into mRNA. Expression data can be used to infer gene regulation: one might compare microarray data from a wide variety of states of an organism to form hypotheses about the genes involved in each state. In a single-cell organism, one might compare stages of the cell cycle, along with various stress conditions (heat shock, starvation, etc.). One can then apply clustering algorithms to that expression data to determine which genes are co-expressed. For example, the upstream regions (promoters) of co-expressed genes can be searched for over-represented regulatory elements.

Analysis of Protein Expression

Protein microarrays and high throughput (HT) mass spectrometry (MS) can provide a snapshot of the proteins present in a biological sample. Bioinformatics is very much involved in making sense of protein microarray and HT MS data; the former approach faces similar problems as with microarrays targeted at mRNA, the latter involves the problem of matching large amounts of mass data against predicted masses from protein sequence databases, and the complicated statistical analysis of samples where multiple, but incomplete peptides from each protein are detected.

Analysis of Mutations in Cancer

In cancer, the genomes of affected cells are rearranged in complex or even unpredictable ways. Massive sequencing efforts are used to identify previously unknown point mutations in a variety of genes in cancer. Bioinformaticians continue to produce specialized automated systems to manage the sheer volume of sequence data produced, and they create new algorithms and software to compare the sequencing results to the growing collection of human genome sequences and germline polymorphisms.

New physical detection technologies are employed, such as oligonucleotidem icroarrays to identify chromosomal gains and losses (called comparative genomic hybridization), and single-nucleotide polymorphism arrays to detect known point mutations. These detection methods simultaneously measure several hundred thousand sites throughout the genome, and when used in high-throughput to measure thousands of samples, generate terabytes of data per experiment. Again the massive amounts and new types of data generate new opportunities for bioinformaticians.

The data is often found to contain considerable variability, or noise, and thus Hidden Markov model and change-point analysis methods are being developed to infer real copy number changes.

Another type of data that requires novel informatics development is the analysis of lesions found to be recurrent among many tumors.

Comparative Genomics

The core of comparative genome analysis is the establishment of the correspondence between genes (orthology analysis) or other genomic features in different organisms. It is these intergenomic maps that make it possible to trace the evolutionary processes responsible for the divergence of two genomes. A multitude of evolutionary events acting at various organizational levels shape genome evolution. At the lowest level, point mutations affect individual nucleotides. At a higher level, large chromosomal segments undergo duplication, lateral transfer, inversion, transposition, deletion and insertion. Ultimately, whole genomes are involved in processes of hybridization, polyploidization and endosymbiosis, often leading to rapid speciation.

The complexity of genome evolution poses many exciting challenges to developers of mathematical models and algorithms, who have recourse to a spectra of algorithmic, statistical and mathematical techniques, ranging from exact, heuristics, fixed parameter and approximation algorithms for problems based on parsimony models to Markov Chain Monte Carlo algorithms for Bayesian analysis of problems based on probabilistic models.

Many of these studies are based on the homology detection and protein families computation.

Modelling Biological Systems

Systems biology involves the use of computer simulations of cellular subsystems (such as the networks of metabolites and enzymes which comprise metabolism, signal transduction pathways and gene regulatory networks) to both analyse and visualize the complex connections of these cellular processes. Artificial life or virtual evolution attempts to understand evolutionary processes via the computer simulation of simple (artificial) life forms.

High-throughput Image Analysis

Computational technologies are used to accelerate or fully automate the processing, quantification and analysis of large amounts of high-information-content biomedical imagery. Modern image analysis systems augment an observer's ability to make measurements from a large or complex set of images, by improving accuracy, objectivity, or speed.

A fully developed analysis system may completely replace the observer. Although these systems are not unique to biomedical imagery, biomedical imaging is becoming more important for both diagnostics and research. Some examples are:

- high-throughput and high-fidelity quantification and subcellular localization (high-content screening, cytohistopathology, Bioimage informatics)
- morphometrics
- clinical image analysis and visualization
- determining the real-time air-flow patterns in breathing lungs of living animals
- quantifying occlusion size in real-time imagery from the development of and recovery during arterial injury

- making behavioural observations from extended video recordings of laboratory animals
- infrared measurements for metabolic activity determination
- inferring clone overlaps in DNA mapping, e.g. the Sulston score

Structural Bioinformatic Approaches

Prediction of Protein Structure

Protein structure prediction is another important application of bioinformatics. The amino acid sequence of a protein, the so-called primary structure, can be easily determined from the sequence on the gene that codes for it.

In the vast majority of cases, this primary structure uniquely determines a structure in its native environment. (Of course, there are exceptions, such as the bovine spongi form encephalopathy – a.k.a. Mad Cow Disease – prion.) Knowledge of this structure is vital in understanding the function of the protein. For lack of better terms, structural information is usually classified as one of secondary, tertiary and quaternary structure. A viable general solution to such predictions remains an open problem. As of now, most efforts have been directed towards heuristics that work most of the time.

One of the key ideas in bioinformatics is the notion of homology. In the genomic branch of bioinformatics, homology is used to predict the function of a gene: if the sequence of gene A, whose function is known, is homologous to the sequence of gene B, whose function is unknown, one could infer that B may share A's function.

In the structural branch of bioinformatics, homology is used to determine which parts of a protein are important in structure formation and interaction with other proteins. In a technique called homology modelling, this information is used to predict the structure of a protein once the structure of a homologous protein is known. This currently remains the only way to predict protein structures reliably.

One example of this is the similar protein homology between hemoglobin in humans and the hemoglobin in legumes (leghemoglobin). Both serve the same purpose of transporting oxygen in the organism.

Though both of these proteins have completely different amino acid sequences, their protein structures are virtually identical, which reflects their near identical purposes.

Other techniques for predicting protein structure include protein threading and de novo (from scratch) physics-based modelling.

Molecular Interaction

Efficient software is available today for studying interactions among proteins, ligands and peptides. Types of interactions most often encountered in the field include – Protein–ligand (including drug), protein–protein and protein–peptide.

Molecular dynamic simulation of movement of atoms about rotatable bonds is the fundamental principle behind computational algorithms, termed docking algorithms for studying molecular interactions.

Docking Algorithms

In the last two decades, tens of thousands of protein three-dimensional structures have been determined by X-ray crystallography and Protein nuclear magnetic resonance spectroscopy (protein NMR). One central question for the biological scientist is whether it is practical to predict possible protein–protein interactions only based on these 3D shapes, without doing protein–protein interaction experiments. A variety of methods have been developed to tackle the Protein–protein docking problem, though it seems that there is still much work to be done in this field.

Software and Tools

Software tools for bioinformatics range from simple command-line tools, to more complex graphical programs and standalone web-services available from various bioinformatics companies or public institutions.

Open Source Bioinformatics Software

Many free and open source software tools have existed and continued to grow since the 1980s. The combination of a continued need for new algorithms for the analysis of emerging types of biological readouts, the potential for innovative in silico experiments, and freely available open code bases have helped to create opportunities for all research groups to contribute to both bioinformatics and the range of open source software available, regardless of their funding arrangements. The open source tools often act as incubators of ideas, or community-supported plug-ins in commercial applications. They may also provide de facto standards and shared object models for assisting with the challenge of bioinformation integration.

The range of open source software packages includes titles such as Bioconductor, BioPerl, BioJava,BioRuby, Bioclipse, EMBOSS, Taverna workbench, and UGENE. In order to maintain this tradition and create further opportunities, the non-profit Open Bioinformatics Foundation have supported the annual Bioinformatics Open Source Conference (BOSC) since 2000.

Web Services in Bioinformatics

SOAP and REST-based interfaces have been developed for a wide variety of bioinformatics applications allowing an application running on one computer in one part of the world to use algorithms, data and computing resources on servers in other parts of the world. The main advantages derive from the fact that end users do not have to deal with software and database maintenance overheads.

Basic bioinformatics services are classified by the EBI into three categories: SSS (Sequence Search Services), MSA (Multiple Sequence Alignment) and BSA (Biological Sequence Analysis). The availability of these service-oriented bioinformatics resources demonstrate the applicability of web based bioinformatics solutions, and range from a collection of standalone tools with a common data format under a single, standalone or web-based interface, to integrative, distributed and extensible bioinformatics workflow management systems.

Cryptanalysis

Figure: *Close-up of the rotors in a Fialka cipher machine*

Cryptanalysis (from the Greek kryptós, "hidden", andanalýein, "to loosen" or "to untie") is the study of methods for obtaining the meaning of encrypted information, without access to the secret information that is normally required to do so. Typically, this involves knowing how the system works and finding a secret key. In non-technical language, this is the practice of codebreaking or cracking the code, although these phrases also have a specialised technical meaning.

Cryptanalysis is also used to refer to any attempt to circumvent the security of other types of cryptographic algorithms and protocols in general, and not justencryption. However, cryptanalysis usually excludes methods of attack that do not primarily target weaknesses in the actual cryptography, such as bribery, physical coercion, burglary, keystroke logging, and social engineering, although these types of attack are an important concern and are often more effective than traditional cryptanalysis. Even though the goal has been the same, the methods and techniques of cryptanalysis have changed drastically through the history of cryptography, adapting to increasing cryptographic complexity, ranging from the pen-and-paper methods of the past, through machines like Bombes and Colossus computers at Bletchley Park in World War II, to the computer-based schemes of the present.

The results of cryptanalysis have also changed — it is no longer possible to have unlimited success in codebreaking, and there is a hierarchical classification of what constitutes an attack. In the mid-1970s, a new class of cryptography was introduced: asymmetric cryptography. Methods for breaking these cryptosystems are typically radically different from before, and usually involve solving carefully constructed problems in pure mathematics, the best-known being integer factorization.

History of Cryptanalysis

Cryptanalysis has coevolved together with cryptography, and the contest can be traced through the history of cryptography—new ciphers being designed to replace old broken designs, and new cryptanalytic techniques invented to crack the improved schemes. In practice, they are viewed as two sides of the same coin: in order to create secure cryptography, you have to design against possible cryptanalysis.

Classical Cryptanalysis

Although the actual word "cryptanalysis" is relatively recent (it was coined by William Friedman in 1920), methods for breaking codes and ciphers are much older. The first known recorded explanation of cryptanalysis was given by 9th-century Arabian polymath, Al-Kindi (also known as "Alkindus" in Europe), in A Manuscript on Deciphering Cryptographic Messages. This treatise includes a description of the method of frequency analysis (Ibrahim Al-Kadi, 1992- ref-3). Italian scholar Giambattista della Porta was author of a seminal work on cryptanalysis "De Furtivis Literarum Notis".

Frequency analysis is the basic tool for breaking mostclassical ciphers. In natural languages, certain letters of the alphabet appear more frequently than others; in English, "E" is likely to be the most common letter in any sample of plaintext. Similarly, the digraph "TH" is the most likely pair of letters in English, and so on. Frequency analysis relies on a cipher failing to hide these statistics.

For example, in a simple substitution cipher (where each letter is simply replaced with another), the most frequent letter in the ciphertext would be a likely candidate for "E". Frequency analysis of such a cipher is therefore relatively easy, provided that the ciphertext is long enough to give a reasonably representative count of the letters of the alphabet that it contains.

In Europe during the 15th and 16th centuries, the idea of a polyalphabetic substitution cipher was developed, among others by the French diplomat Blaise de Vigenère (1523–96). For some three centuries, the Vigenère cipher, which uses a repeating key to select different encryption alphabets in rotation, was considered to be completely secure (le chiffre indéchiffrable—"the indecipherable cipher"). Nevertheless, Charles Babbage (1791–1871) and later, independently, Friedrich Kasiski(1805–81) succeeded in breaking this cipher. During World War I, inventors in several countries developed rotor cipher machines such as Arthur Scherbius' Enigma, in an attempt to minimise the repetition that had been exploited to break the Vigenère system.

In practice, frequency analysis relies as much on linguistic knowledge as it does on statistics, but as ciphers became more complex, mathematics became more important in cryptanalysis. This change was particularly evident before and during World War II, where efforts to crack Axis ciphers required new levels of mathematical sophistication. Moreover, automation was first applied to cryptanalysis in that era with the Polish Bomba device, the British Bombe development of it, the use of punched card equipment, and in the Colossus computers — the first electronic digital computers to be controlled by a program.

Depth

Sending two or more messages with the same key is an insecure process. To a cryptanalyst the messages are then said to be "in depth". This may be detected by the messages having the same indicator by which the sending operator informs the receiving operator about the key generator initial settings for the message.

In a symmetrical cipher, the same key that was applied to the plaintext to produce the ciphertext is applied to the ciphertext to recover the plaintext. A simple example of such a system is the Vernam cipher in which a long key is bit-for-bit combined with the plaintext or ciphertext using the "XOR" operator (symbolized by $\otimes$):

$$Plaintext \oplus Key = Ciphertext \text{ at each bit position}$$

and

$$Ciphertext \oplus Key = Plaintext \text{ at each bit position}$$

When the ciphertexts are in depth, combining them eliminates the common key, leaving just a combination of the two plaintexts:

$$Ciphertext1 \oplus Ciphertext2 = \alpha = Plaintext1 \oplus Plaintext2$$ at each bit position.

The individual plaintexts can then be worked out by trying probable words (or phrases) at various locations; when the corresponding section of á is XORed with a correct guess at the probable word, a stretch of the other plaintext is revealed, which can be recognized and (quite often) extended at each end, working back and forth between the plaintexts to recover much or all of them.

When a recovered plaintext is then combined with its ciphertext, the key is revealed:

$$Plaintext1 \oplus Ciphertext1 = Key$$ at each bit position

Knowledge of keys from a cipher may allow cryptanalysts to work out the system used for constructing them.

Depth reading can be applied in more complicated contexts to eliminate some of a system's parameters, or to "stack" messages so that source language statistics become exploitable. One way to properly align multiple messages that use the same periodic key starting at different positions is by the kappa test.

Modern Cryptanalysis

Figure: *The Bombe replicated the action of several Enigma machines wired together. Each of the rapidly rotating drums, pictured above in a Bletchley Park museum mockup, simulated the action of an Enigma rotor.*

Even though computation was used to great effect in cryptanalysis of the Enigma and other systems during World War II, it also made possible new methods of cryptography orders of magnitude more complex than ever before. Taken as a whole, modern cryptography has become much more impervious to cryptanalysis than the pen-and-paper systems of the past, and now seems to have the upper hand against pure cryptanalysis. The historian David Kahn notes,

"Many are the cryptosystems offered by the hundreds of commercial vendors today that cannot be broken by any known methods of cryptanalysis. Indeed, in such systems even a chosen plaintext attack, in which a selected plaintext is matched against its ciphertext, cannot yield the key that unlock[s] other messages. In a sense, then, cryptanalysis is dead.

But that is not the end of the story. Cryptanalysis may be dead, but there is - to mix my metaphors - more than one way to skin a cat.". Kahn goes on to mention increased opportunities for interception, bugging, side channel attacks, and quantum computers as replacements for the traditional means of cryptanalysis. In 2010, former NSA technical director Brian Snow said that both academic and government cryptographers are "moving very slowly forward in a mature field."

However, any postmortems for cryptanalysis may be premature. While the effectiveness of cryptanalytic methods employed by intelligence agencies remains unknown, many serious attacks against both academic and practical cryptographic primitives have been published in the modern era of computer cryptography:

- The block cipher Madryga, proposed in 1984 but not widely used, was found to be susceptible to ciphertext-only attacks in 1998.
- FEAL-4, proposed as a replacement for the DES standard encryption algorithm but not widely used, was demolished by a spate of attacks from the academic community, many of which are entirely practical.
- The A5/1, A5/2, CMEA, and DECT systems used in mobile and wireless phone technology can all be broken in hours, minutes or even in real-time using widely-available computing equipment.
- Brute-force keyspace search has broken some real-world ciphers and applications, including single-DES, 40-bit "export-strength" cryptography, and the DVD Content Scrambling System.
- In 2001, Wired Equivalent Privacy (WEP), a protocol used to secure Wi-Fi wireless networks, was shown to be breakable in practice because of a weakness in the RC4 cipher and aspects of the WEP design that made related-key attacks practical. WEP was later replaced by Wi-Fi Protected Access.
- In 2008, researchers conducted a proof-of-concept break of SSL using weaknesses in the MD5hash function and certificate issuer practices that made it possible to exploit collision attacks on hash functions. The certificate issuers involved changed their practices to prevent the attack from being repeated.

Thus, while the best modern ciphers may be far more resistant to cryptanalysis than the Enigma, cryptanalysis and the broader field of information security remain quite active.

The Results of Cryptanalysis

Successful cryptanalysis has undoubtedly influenced history; the ability to read the presumed-secret thoughts and plans of others can be a decisive advantage.

For example, in England in 1587, Mary, Queen of Scots was tried and executed for treason for her involvement in three plots to assassinate Elizabeth I of England which were known about because her coded correspondence with fellow conspirators had been deciphered by Thomas Phelippes; in World War I, the breaking of the Zimmermann Telegram was instrumental in bringing the United States into the war; in World War II, the cryptanalysis of the German ciphers — including the Enigma machine and the Lorenz cipher — has been credited with everything between shortening the end of the European war by a few months to determining the eventual result. The United States also benefited from the cryptanalysis of the Japanese Purple code.

Governments have long recognized the potential benefits of cryptanalysis for intelligence, both military and diplomatic, and established dedicated organizations devoted to breaking the codes and ciphers of other nations, for example, GCHQ and the NSA, organizations which are still very active today.

In 2004, it was reported that the United States had broken Iranian ciphers. (It is unknown, however, whether this was pure cryptanalysis, or whether other factors were involved:).

Types of Cryptanalytic Attack

Cryptanalytic attacks vary in potency and how much of a threat they pose to real-world cryptosystems. A certificational weakness is a theoretical attack that is unlikely to be applicable in any real-world situation; the majority of results found in modern cryptanalytic research are of this type. Essentially, the practical importance of an attack is dependent on the answers to the following four questions:

1. What knowledge and capabilities does the attacker need?
2. How much additional secret information is deduced?
3. How much computation is required? (What is the computational complexity?)
4. Does the attack break the full cryptosystem, or only a weakened version?

Access Needed for the Attack

Cryptanalysis can be performed under a number of assumptions about how much access the attacker has to the system under attack. As a basic starting point it is normally assumed that, for the purposes of analysis, the general algorithm is known; this is Kerckhoffs' principle of "the enemy knows the system".

This is a reasonable assumption in practice — throughout history, there are countless examples of secret algorithms falling into wider knowledge, variously through espionage, betrayal and reverse engineering. (On occasion, ciphers have been reconstructed through pure deduction; for example, the German Lorenz cipher and the Japanese Purple code, and a variety of classical schemes).

Other assumptions include:

- Ciphertext-only: the cryptanalyst has access only to a collection of ciphertexts or codetexts.
- Known-plaintext: the attacker has a set of ciphertexts to which he knows the corresponding plaintext.
- Chosen-plaintext (chosen-ciphertext): the attacker can obtain the ciphertexts (plaintexts) corresponding to an arbitrary set of plaintexts (ciphertexts) of his own choosing.
- Adaptive chosen-plaintext: like a chosen-plaintext attack, except the attacker can choose subsequent plaintexts based on information learned from previous encryptions. Similarly Adaptive chosen ciphertext attack.
- Related-key attack: Like a chosen-plaintext attack, except the attacker can obtain ciphertexts encrypted under two different keys. The keys are unknown, but the relationship between them is known; for example, two keys that differ in the one bit.

These types of attack clearly differ in how plausible they would be to mount in practice. Although some are more likely than others, cryptographers will often take a conservative approach to security and assume the worst-case when designing algorithms, reasoning that if a scheme is secure even against unrealistic threats, then it should also resist real-world cryptanalysis as well. The assumptions are often more realistic than they might seem upon first glance. For a known-plaintext attack, the cryptanalyst might well know or be able to guess at a likely part of the plaintext, such as an encrypted letter beginning with "Dear Sir", or a computer session starting with "LOGIN:". A chosen-plaintext attack is less likely, but it is sometimes plausible: for example, you could convince someone to forward a message you have given them, but in encrypted form. Related-key attacks are mostly theoretical, although they can be realistic in certain situations, for example, when constructing cryptographic hash functions using a block cipher.

Usefulness of Attack Results

The results of cryptanalysis can also vary in usefulness. For example, cryptographer Lars Knudsen(1998) classified various types of attack on block ciphers according to the amount and quality of secret information that was discovered:

- Total break — the attacker deduces the secret key.
- Global deduction — the attacker discovers a functionally equivalent algorithm for encryption and decryption, but without learning the key.
- Instance (local) deduction — the attacker discovers additional plaintexts (or ciphertexts) not previously known.
- Information deduction — the attacker gains some Shannon information about plaintexts (or ciphertexts) not previously known.
- Distinguishing algorithm — the attacker can distinguish the cipher from a random permutation.

Similar considerations apply to attacks on other types of cryptographic algorithm.

Computational Resources Required

Attacks can also be characterised by the resources they require. Those resources include:

- Time — the number of computation steps (like encryptions) which must be performed.
- Memory — the amount of storage required to perform the attack.
- Data — the quantity of plaintexts and ciphertexts required.

It's sometimes difficult to predict these quantities precisely, especially when the attack isn't practical to actually implement for testing. But academic cryptanalysts tend to provide at least the estimated order of magnitude of their attacks' difficulty, saying, for example, "SHA-1 collisions now 2". Bruce Schneier notes that even computationally impractical attacks can be considered breaks: "Breaking a cipher simply means finding a weakness in the cipher that can be exploited with a complexity less than brute force.

Never mind that brute-force might require 2 encryptions; an attack requiring 2 encryptions would be considered a break...simply put, a break can just be a certificational weakness: evidence that the cipher does not perform as advertised."

Partial Breaks

Academic attacks are often against weakened versions of a cryptosystem, such as a block cipher or hash function with some rounds removed. Many, but not all, attacks become exponentially more difficult to execute as rounds are added to a cryptosystem, so it's possible for the full cryptosystem to be strong even though reduced-round variants are weak. Nonetheless, partial breaks that come close to breaking the original cryptosystem may mean that a full break will follow; the successful attacks on DES, MD5, and SHA-1 were all preceded by attacks on weakened versions.

Academic Weakness versus Practical Weakness

In academic cryptography, a weakness or a break in a scheme is usually defined quite conservatively: it might require impractical amounts of time, memory, or known plaintexts.

It also might require the attacker be able to do things many real-world attackers can't: for example, the attacker may need to choose particular plaintexts to be encrypted or even to ask for plaintexts to be encrypted using several keys related to the secret key. Furthermore, it might only reveal a small amount of information, enough to prove the cryptosystem imperfect but too little to be useful to real-world attackers. Finally, an attack might only apply to a weakened version of cryptographic tools, like a reduced-round block cipher, as a step towards breaking of the full system.

Cryptanalysis of Asymmetric Cryptography

Asymmetric cryptography (or public key cryptography) is cryptography that relies on using two keys; one private, and one public. Such ciphers invariably rely on "hard" mathematical problems as the basis of their security, so an obvious point of attack is to develop methods for solving the problem. The security of two-key cryptography depends on mathematical questions in a way that single-key cryptography generally does not, and conversely links cryptanalysis to wider mathematical research in a new way.

Asymmetric schemes are designed around the (conjectured) difficulty of solving various mathematical problems. If an improved algorithm can be found to solve the problem, then the system is weakened. For example, the security of the Diffie-Hellman key exchange scheme depends on the difficulty of calculating the discrete logarithm. In 1983, Don Coppersmith found a faster way to find discrete logarithms (in certain groups), and thereby requiring cryptographers to use larger groups (or different types of groups). RSA's security depends (in part) upon the difficulty of integer factorization — a breakthrough in factoring would impact the security of RSA. In 1980, one could factor a difficult 50-digit number at an expense of 10 elementary computer operations. By 1984 the state of the art in factoring algorithms had advanced to a point where a 75-digit number could be factored in 10 operations. Advances in computing technology also meant that the operations could be performed much faster, too. Moore's law predicts that computer speeds will continue to increase.

Factoring techniques may continue to do so as well, but will most likely depend on mathematical insight and creativity, neither of which has ever been successfully predictable. 150-digit numbers of the kind once used in RSA have been factored.

The effort was greater than above, but was not unreasonable on fast modern computers. By the start of the 21st century, 150-digit numbers were no longer considered a large enough key size for RSA. Numbers with several hundred digits were still considered too hard to factor in 2005, though methods will probably continue to improve over time, requiring key size to keep pace or other methods such as elliptic curve cryptography to be used.

Another distinguishing feature of asymmetric schemes is that, unlike attacks on symmetric cryptosystems, any cryptanalysis has the opportunity to make use of knowledge gained from the public key.

Quantum Computing Applications for Cryptanalysis

Quantum computers, which are still in the early phases of research, have potential use in cryptanalysis. For example, Shor's Algorithm could factor large numbers in polynomial time, in effect breaking some commonly used forms of public-key encryption. By using Grover's algorithm on a quantum computer, brute-force key search can be made quadratically faster. However, this could be countered by doubling the key length.

Pop Music

Pop music (a term that originally derives from an abbreviation of "popular") is usually understood to be commercially recorded music, often oriented toward a youth market, usually consisting of relatively short, simple songs utilizing technological innovations to produce new variations on existing themes.

Definitions

David Hatch and Stephen Millward define pop music as "a body of music which is distinguishable from popular, jazz and folk musics". Although pop music is often seen as oriented towards the singles charts it is not the sum of all chart music, which has always contained songs from a variety of sources, including classical, jazz, rock, and novelty songs, while pop music as a genre is usually seen as existing and developing separately. Thus "pop music" may be used to describe a distinct genre, aimed at a youth market, often characterized as a softer alternative to rock and roll.

Origin of the Term

The term "pop song" is first recorded as being used in 1926, in the sense of a piece of music "having popular appeal". Hatch and Millward indicate that many events in the history of recording in the 1920s can be seen as the birth of the modern pop music industry, including in country, blues and hillbilly music. According to Grove Music Online, the term "pop music" "originated in Britain in the mid-1950s as a description for Rock and roll and the new youth music styles that it influenced...".

The Oxford Dictionary of Music states that while pop's "earlier meaning meant concerts appealing to a wide audience...[,] since the late 1950s, however, pop has had the special meaning of non-classical music, usually in the form of songs, performed by such artists as the Beatles, the Rolling Stones, ABBA, etc." Grove Music Online also states that "... in the early 1960s [the term] 'pop music' competed terminologically with Beat music [in England], while in the USA its coverage overlapped (as it still does) with that of 'rock and roll'." Chambers' Dictionary mentions the contemporary usage of the term "pop art"; Grove Music Online states that the "term pop music... seems to have been a spin-off from the terms pop art and pop culture, coined slightly earlier, and referring to a whole range of new, often American, media-culture products".

From about 1967 the term was increasingly used in opposition to the term rock music, a division that gave generic significance to both terms. Whereas rock aspired to authenticity and an expansion of the possibilities of popular music, pop was more commercial, ephemeral and accessible. According to Simon Frith pop music is produced "as a matter of enterprise not art", is "designed to appeal to everyone" and "doesn't come from any particular place or mark off any particular taste".

It is "not driven by any significant ambition except profit and commercial reward... and, in musical terms, it is essentially conservative". It is, "provided from on high (by record companies, radio programmers and concert promoters) rather than being made from below... Pop is not a do-it-yourself music but is professionally produced and packaged".

Influences and Development

Throughout its development, pop music has absorbed influences from most other genres of popular music. Early pop music drew on the sentimental ballad for its form, gained its use of vocal harmonies from gospel and soul music, instrumentation from jazz, country, and rock music, orchestration from classical music, tempo from dance music, backing from electronic music, rhythmic elements from hip-hop music, and has recently appropriated spoken passages from rap.

It has also made use of technological innovation. In the 1940s improved microphone design allowed a more intimate singing style and ten or twenty years later inexpensive and more durable 45 r.p.m. records for singles "revolutionized the manner in which pop has been disseminated" and helped to move pop music to 'a record/radio/film star system'. Another technological change was the widespread availability of television in the 1950s; with televised performances, "pop stars had to have a visual presence".

In the 1960s, the introduction of inexpensive, portable transistor radios meant that teenagers could listen to music outside of the home. Multi-track recording (from the 1960s); and digital sampling (from the 1980s) have also been utilized as methods for the creation and elaboration of pop music. By the early 1980s, the promotion of pop music had been greatly affected by the rise of Music Television channels like MTV, which "favoured those artists such as Michael Jackson andMadonna who had a strong visual appeal".

Pop music has been dominated by the American and (from the mid-1960s) British music industries, whose influence has made pop music something of an international monoculture, but most regions and countries have their own form of pop music, sometimes producing local versions of wider trends, and lending them local characteristics. Some of these trends (for example Europop) have had a significant impact of the development of the genre.

According to Grove Music Online, "Western-derived pop styles, whether coexisting with or marginalizing distinctively local genres, have spread throughout the world and have come to constitute stylistic common denominators in global commercial music cultures". Some non-Western countries, such as Japan, have developed a thriving pop music industry, most of which is devoted to Western-style pop, has for several years has produced a greater quantity of music of everywhere except the USA. The spread of Western-style pop music has been interpreted variously as representing processes of Americanization, homogenization, modernization, creative appropriation, cultural imperialism, and/or a more general process of globalization.

Characteristics

Musicologists often identify the following characteristics as typical of the pop music genre:

- an aim of appealing to a general audience, rather than to a particular subculture or ideology

- an emphasis on craftsmanship rather than formal "artistic" qualities
- an emphasis on recording, production, and technology, over live performance
- a tendency to reflect existing trends rather than progressive developments
- much pop music is intended to encourage dancing, or it uses dance-oriented beats or rhythms

The main medium of pop music is the song, often between two and a half and three and a half minutes in length, generally marked by a consistent and noticeable rhythmic element, a mainstream style and a simple traditional structure. Common variants include the verse-chorus form and the thirty-two-bar form, with a focus on melodies and catchy hooks, and a chorus that contrasts melodically, rhythmically and harmonically with the verse. The beat and the melodies tend to be simple, with limited harmonic accompaniment. The lyrics of modern pop songs typically focus on simple themes – often love and romantic relationships – although there are notable exceptions.

Harmony in pop music is often "that of classical European tonality, only more simple-minded. "Clichés include the barbershop harmony (i.e. moving from a secondary dominant harmony to a dominant harmony, and then to the tonic) and blues scale-influenced harmony. "The influence of the circle-of-fifths paradigm has declined since the mid-1950s.

The harmonic languages of rock and soul have moved away from the all-encompassing influence of the dominant function....There are other tendencies (perhaps also traceable to the use of a guitar as a composing instrument) – pedal-pointharmonies, root motion by diatonic step, modal harmonic and melodic organization – that point away from functional tonality and toward a tonal sense that is less directional, more free-floating."

Travelling Salesman Problem

The travelling salesman problem (TSP) is an NP-hard problem in combinatorial optimisation studied in operations research and theoretical computer science. Given a list of cities and their pairwise distances, the task is to find the shortest possible tour that visits each city exactly once. It is a special case of the Travelling purchaser problem. The problem was first formulated as a mathematical problem in 1930 and is one of the most intensively studied problems in optimisation. It is used as a benchmark for many optimisation methods. Even though the problem is computationally difficult, a large number of heuristics and exact methods are known, so that some instances with tens of thousands of cities can be solved.

The TSP has several applications even in its purest formulation, such as planning, logistics, and the manufacture of microchips. Slightly modified, it appears as a sub-problem in many areas, such as DNA sequencing. In these applications, the concept city represents, for example, customers, soldering points, or DNA fragments, and the concept distance represents travelling times or cost, or a similarity measure between DNA fragments.

In many applications, additional constraints such as limited resources or time windows make the problem considerably harder. In the theory of computational complexity, the decision version of the TSP (where, given a length L, the task is to decide whether any tour is shorter than L) belongs to the class of NP-complete problems. Thus, it is likely that the worst case running time for any algorithm for the TSP increases exponentially with the number of cities.

History

The origins of the travelling salesman problem are unclear. A handbook for travelling salesmen from 1832 mentions the problem and includes example tours through Germany and Switzerland, but contains no mathematical treatment.

The travelling salesman problem was defined in the 1800s by the Irish mathematician W. R. Hamilton and by the British mathematician Thomas Kirkman. Hamilton's Icosian Game was a recreational puzzle based on finding a Hamiltonian cycle. The general form of the TSP appears to have been first studied by mathematicians during the 1930s in Vienna and at Harvard, notably by Karl Menger, who defines the problem, considers the obvious brute-force algorithm, and observes the non-optimality of the nearest neighbour heuristic:

We denote by messenger problem (since in practice this question should be solved by each postman, anyway also by many travelers) the task to find, for finitely many points whose pairwise distances are known, the shortest route connecting the points. Of course, this problem is solvable by finitely many trials. Rules which would push the number of trials below the number of permutations of the given points, are not known. The rule that one first should go from the starting point to the closest point, then to the point closest to this, etc., in general does not yield the shortest route.

Hassler Whitney at Princeton University introduced the name travelling salesman problem soon after. In the 1950s and 1960s, the problem became increasingly popular in scientific circles in Europe and the USA. Notable contributions were made by George Dantzig, Delbert Ray Fulkerson and Selmer M. Johnson at the RAND Corporation in Santa Monica, who expressed the problem as an integer linear program and developed the cutting plane method for its solution. With these new methods they solved an instance with 49 cities to optimality by constructing a tour and proving that no other tour could be shorter. In the following decades, the problem was studied by many researchers from mathematics, computer science, chemistry, physics, and other sciences.

Richard M. Karp showed in 1972 that the Hamiltonian cycle problem was NP-complete, which implies the NP-hardness of TSP. This supplied a mathematical explanation for the apparent computational difficulty of finding optimal tours.

Great progress was made in the late 1970s and 1980, when Grötschel, Padberg, Rinaldi and others managed to exactly solve instances with up to 2392 cities, using cutting planes and branch-and-bound.

In the 1990s, Applegate, Bixby, Chvátal, and Cook developed the program Concorde that has been used in many recent record solutions. Gerhard Reinelt published the TSPLIB in 1991, a collection of benchmark instances of varying difficulty, which has been used by many research groups for comparing results. In 2005, Cook and others computed an optimal tour through a 33,810-city instance given by a microchip layout problem, currently the largest solved TSPLIB instance. For many other instances with millions of cities, solutions can be found that are guaranteed to be within 1% of an optimal tour.

Description

As a graph problem:

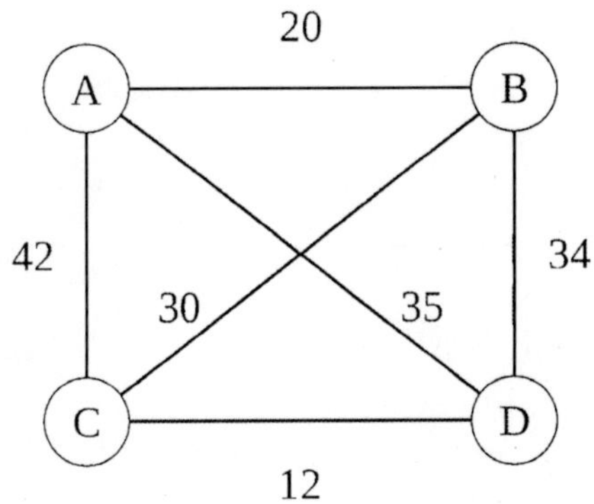

Figure: *Symmetric TSP with four cities*

TSP can be modelled as an undirected weighted graph, such that cities are the graph's vertices, paths are the graph's edges, and a path's distance is the edge's length. A TSP tour becomes a Hamiltonian cycle if and only if every edge has the same distance. Often, the model is a complete graph (i.e., an edge connects each pair of vertices). If no path exists between two cities, adding an arbitrarily long edge will complete the graph without affecting the optimal tour.

Asymmetric and Symmetric

In the symmetric TSP, the distance between two cities is the same in each opposite direction, forming an undirected graph. This symmetry halves the number of possible solutions. In the asymmetric TSP, paths may not exist in both directions or the distances might be different, forming a directed graph.

Traffic collisions, one-way streets, and airfares for cities with different departure and arrival fees are examples of how this symmetry could break down.

Related Problems

- An equivalent formulation in terms of graph theory is: Given a complete weighted graph (where the vertices would represent the cities, the edges would represent the roads, and the weights would be the cost or distance of that road), find a Hamiltonian cycle with the least weight.
- The requirement of returning to the starting city does not change the computational complexity of the problem.
- Another related problem is the bottleneck travelling salesman problem (bottleneck TSP): Find a Hamiltonian cycle in a weighted graph with the minimal weight of the weightiest edge. The problem is of considerable practical importance, apart from evident transportation and logistics areas. A classic example is in printed circuit manufacturing: scheduling of a route of the drill machine to drill holes in a PCB. In robotic machining or drilling applications, the "cities" are parts to machine or holes (of different sizes) to drill, and the "cost of travel" includes time for retooling the robot (single machine job sequencing problem).
- The generalised travelling salesman problem deals with "states" that have (one or more) "cities" and the salesman has to visit exactly one "city" from each "state". Also known as the "travelling politician problem". One application is encountered in ordering a solution to the cutting stock problem in order to minimise knife changes. Another is concerned with drilling in semiconductor manufacturing. Surprisingly, Behzad and Modarres demonstrated that the generalised travelling salesman problem can be transformed into a standard travelling salesman problem with the same number of cities, but a modified distance matrix.
- The sequential ordering problem deals with the problem of visiting a set of cities where precedence relations between the cities exist.

- The travelling purchaser problem deals with a purchaser who is charged with purchasing a set of products. He can purchase these products in several cities, but at different prices and not all cities offer the same products. The objective is to find a route between a subset of the cities, which minimizes total cost (travel cost + purchasing cost) and which enables the purchase of all required products.

Computing a Solution

The traditional lines of attack for the NP-hard problems are the following:

- Devising algorithms for finding exact solutions (they will work reasonably fast only for small problem sizes).
- Devising "suboptimal" or heuristic algorithms, i.e., algorithms that deliver either seemingly or probably good solutions, but which could not be proved to be optimal.
- Finding special cases for the problem ("subproblems") for which either better or exact heuristics are possible.

Computational Complexity

The problem has been shown to be NP-hard (more precisely, it is complete for the complexity class FP), and the decision problem version ("given the costs and a number x, decide whether there is a round-trip route cheaper than x") is NP-complete. The bottleneck travelling salesman problem is also NP-hard. The problem remains NP-hard even for the case when the cities are in the plane with Euclidean distances, as well as in a number of other restrictive cases. Removing the condition of visiting each city "only once" does not remove the NP-hardness, since it is easily seen that in the planar case there is an optimal tour that visits each city only once (otherwise, by the triangle inequality, a shortcut that skips a repeated visit would not increase the tour length).

Complexity of Approximation

In the general case, finding a shortest travelling salesman tour is NPO-complete. If the distance measure is a metric and symmetric, the problem becomes APX-complete and Christofides's algorithm approximates it within 1.5. If the distances are restricted to 1 and 2 (but still are a metric) the approximation ratio becomes 7/6. In the asymmetric, metric case, only logarithmic performance guarantees are known, the best current algorithm achieves performance ratio 0.814 log n; it is an open question if a constant factor approximation exists. The corresponding maximization problem of finding the longest travelling salesman tour is approximable within 63/38. If the distance function is symmetric, the longest tour can be approximated within 4/3 by a deterministic algorithm and within by a randomised algorithm.

Exact Algorithms

The most direct solution would be to try all permutations (ordered combinations) and see which one is cheapest (using brute force search). The running time for this approach lies within a polynomial factor of O(n!), the factorial of the number of cities, so this solution becomes impractical even for only 20 cities. One of the earliest applications of dynamic programming is the Held–Karp algorithm that solves the problem in time O(n2).

The dynamic programming solution requires exponential space. Using inclusion–exclusion, the problem can be solved in time within a polynomial factor of 2 and polynomial space. Improving these time bounds seems to be difficult. For example, it has not been determined whether an exact algorithm for TSP that runs in time O(1.9999) exists. Other approaches include:

- Various branch-and-bound algorithms, which can be used to process TSPs containing 40–60 cities.
- Progressive improvement algorithms which use techniques reminiscent of linear programming. Works well for up to 200 cities.
- Implementations of branch-and-bound and problem-specific cut generation; this is the method of choice for solving large instances. This approach holds the current record, solving an instance with 85,900 cities.

An exact solution for 15,112 German towns from TSPLIB was found in 2001 using the cutting-plane method proposed by George Dantzig, Ray Fulkerson, and Selmer M. Johnson in 1954, based on linear programming. The computations were performed on a network of 110 processors located at Rice University and Princeton University.

The total computation time was equivalent to 22.6 years on a single 500 MHz Alpha processor. In May 2004, the travelling salesman problem of visiting all 24,978 towns in Sweden was solved: a tour of length approximately 72,500 kilometres was found and it was proven that no shorter tour exists. In March 2005, the travelling salesman problem of visiting all 33,810 points in a circuit board was solved using Concorde TSP Solver: a tour of length 66,048,945 units was found and it was proven that no shorter tour exists. The computation took approximately 15.7 CPU-years (Cook et al. 2006). In April 2006 an instance with 85,900 points was solved using Concorde TSP Solver, taking over 136 CPU-years.

Heuristic and Approximation Algorithms

Various heuristics and approximation algorithms, which quickly yield good solutions have been devised. Modern methods can find solutions for extremely large problems (millions of cities) within a reasonable time which are with a high probability just 2–3% away from the optimal solution. Several categories of heuristics are recognized.

Constructive Heuristics

The nearest neighbour (NN) algorithm (or so-called greedy algorithm) lets the salesman choose the nearest unvisited city as his next move. This algorithm quickly yields an effectively short route. For N cities randomly distributed on a plane, the algorithm on average yields a path 25% longer than the shortest possible path.

However, there exist many specially arranged city distributions which make the NN algorithm give the worst route (Gutin, Yeo, and Zverovich, 2002). This is true for both asymmetric and symmetric TSPs (Gutin and Yeo, 2007). Rosenkrantz et al. [1977] showed that the NN algorithm has the approximation factor $\Theta(\log |V|)$ for instances satisfying the triangle inequality. Constructions based on a minimum spanning tree have an approximation ratio of 2.

The Christofides algorithm achieves a ratio of 1.5. The bitonic tour of a set of points is the minimum-perimetre monotone polygon that has the points as its vertices; it can be computed efficiently by dynamic programming.

Another constructive heuristic, Match Twice and Stitch (MTS) (Kahng, Reda 2004), performs two sequential matchings, where the second matching is executed after deleting all the edges of the first matching, to yield a set of cycles. The cycles are then stitched to produce the final tour.

Iterative Improvement

Pairwise Exchange, or Lin–Kernighan Heuristics

The pairwise exchange or 2-opt technique involves iteratively removing two edges and replacing these with two different edges that reconnect the fragments created by edge removal into a new and shorter tour. This is a special case of the k-opt method. Note that the label Lin–Kernighan is an often heard misnomer for 2-opt. Lin–Kernighan is actually a more general method.

k-opt Heuristic

Take a given tour and delete k mutually disjoint edges. Reassemble the remaining fragments into a tour, leaving no disjoint subtours (that is, don't connect a fragment's endpoints together).

This in effect simplifies the TSP under consideration into a much simpler problem. Each fragment endpoint can be connected to 2k " 2 other possibilities: of 2k total fragment endpoints available, the two endpoints of the fragment under consideration are disallowed.

Such a constrained 2k-city TSP can then be solved with brute force methods to find the least-cost recombination of the original fragments. The k-opt technique is a special case of the V-opt or variable-opt technique.

The most popular of the k-opt methods are 3-opt, and these were introduced by Shen Lin of Bell Labs in 1965. There is a special case of 3-opt where the edges are not disjoint (two of the edges are adjacent to one another).

In practice, it is often possible to achieve substantial improvement over 2-opt without the combinatorial cost of the general 3-opt by restricting the 3-changes to this special subset where two of the removed edges are adjacent. This so-called two-and-a-half-opt typically falls roughly midway between 2-opt and 3-opt, both in terms of the quality of tours achieved and the time required to achieve those tours.

V-opt Heuristic

The variable-opt method is related to, and a generalisation of the k-opt method. Whereas the k-opt methods remove a fixed number (k) of edges from the original tour, the variable-opt methods do not fix the size of the edge set to remove. Instead they grow the set as the search process continues.

The best known method in this family is the Lin–Kernighan method (mentioned above as a misnomer for 2-opt). Shen Lin and Brian Kernighan first published their method in 1972, and it was the most reliable heuristic for solving travelling salesman problems for nearly two decades. More advanced variable-opt methods were developed at Bell Labs in the late 1980s by David Johnson and his research team.

These methods (sometimes called Lin–Kernighan–Johnson) build on the Lin–Kernighan method, adding ideas from tabu search and evolutionary computing. The basic Lin–Kernighan technique gives results that are guaranteed to be at least 3-opt. The Lin–Kernighan–Johnson methods compute a Lin–Kernighan tour, and then perturb the tour by what has been described as a mutation that removes at least four edges and reconnecting the tour in a different way, then v-opting the new tour.

The mutation is often enough to move the tour from the local minimum identified by Lin–Kernighan. V-opt methods are widely considered the most powerful heuristics for the problem, and are able to address special cases, such as the Hamilton Cycle Problem and other non-metric TSPs that other heuristics fail on.

For many years Lin–Kernighan–Johnson had identified optimal solutions for all TSPs where an optimal solution was known and had identified the best known solutions for all other TSPs on which the method had been tried.

Randomised Improvement

Optimised Markov chain algorithms which use local searching heuristic sub-algorithms can find a route extremely close to the optimal route for 700 to 800 cities.

TSP is a touchstone for many general heuristics devised for combinatorial optimisation such as genetic algorithms, simulated annealing, Tabu search, ant colony optimisation, river formation dynamics and the cross entropy method.

Ant Colony Optimisation

Artificial intelligence researcher Marco Dorigo described in 1997 a method of heuristically generating "good solutions" to the TSP using a simulation of an ant colony called ACS (Ant Colony System). It models behaviour observed in real ants to find short paths between food sources and their nest, an emergent behaviour resulting from each ant's preference to follow trail pheromones deposited by other ants. ACS sends out a large number of virtual ant agents to explore many possible routes on the map. Each ant probabilistically chooses the next city to visit based on a heuristic combining the distance to the city and the amount of virtual pheromone deposited on the edge to the city. The ants explore, depositing pheromone on each edge that they cross, until they have all completed a tour. At this point the ant which completed the shortest tour deposits virtual pheromone along its complete tour route (global trail updating). The amount of pheromone deposited is inversely proportional to the tour length: the shorter the tour, the more it deposits.

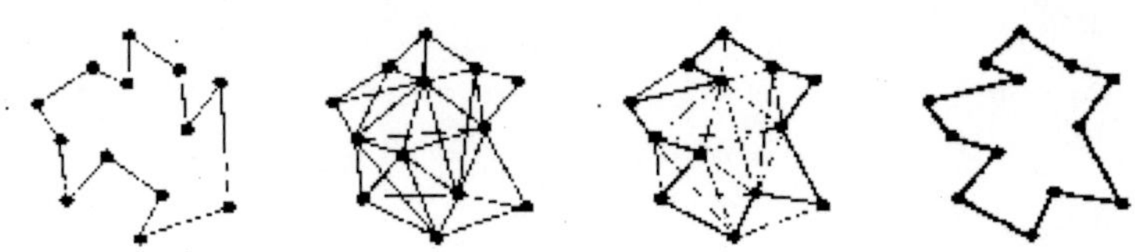

1 2 3 4

Special Cases

Metric TSP

In the metric TSP, also known as delta-TSP or Ä-TSP, the intercity distances satisfy the triangle inequality. A very natural restriction of the TSP is to require that the distances between cities form a metric, i.e., they satisfy the triangle inequality. This can be understood as the absence of "shortcuts", in the sense that the direct connection from A to B is never longer than the route via intermediate C:

$$d_{AB} \leq d_{AC} + d_{CB}$$

The edge lengths then form a metric on the set of vertices. When the cities are viewed as points in the plane, many natural distance functions are metrics, and so many natural instances of TSP satisfy this constraint. The following are some examples of metric TSPs for various metrics.

- In the Euclidean TSP the distance between two cities is the Euclidean distance between the corresponding points.
- In the rectilinear TSP the distance between two cities is the sum of the differences of their x- and y-coordinates. This metric is often called the Manhattan distance or city-block metric.
- In the maximum metric, the distance between two points is the maximum of the absolute values of differences of their x- and y-coordinates.

The last two metrics appear for example in routing a machine that drills a given set of holes in a printed circuit board. The Manhattan metric corresponds to a machine that adjusts first one co-ordinate, and then the other, so the time to move to a new point is the sum of both movements.

The maximum metric corresponds to a machine that adjusts both co-ordinates simultaneously, so the time to move to a new point is the slower of the two movements. In its definition, the TSP does not allow cities to be visited twice, but many applications do not need this constraint. In such cases, a symmetric, non-metric instance can be reduced to a metric one.

This replaces the original graph with a complete graph in which the inter-city distance d_{AB} is replaced by the shortest path between A and B in the original graph.

There is a constant-factor approximation algorithm for the metric TSP due to Christofides that always finds a tour of length at most 1.5 times the shortest tour. In the next paragraphs, we explain a weaker (but simpler) algorithm which finds a tour of length at most twice the shortest tour.

The length of the minimum spanning tree of the network is a natural lower bound for the length of the optimal route. In the TSP with triangle inequality case it is possible to prove upper bounds in terms of the minimum spanning tree and design an algorithm that has a provable upper bound on the length of the route.

The first published (and the simplest) example follows:

1. Construct the minimum spanning tree.
2. Duplicate all its edges. That is, wherever there is an edge from u to v, add a second edge from u to v. This gives us an Eulerian graph.
3. Find a Eulerian cycle in it. Clearly, its length is twice the length of the tree.

4. Convert the Eulerian cycle into the Hamiltonian one in the following way: walk along the Eulerian cycle, and each time you are about to come into an already visited vertex, skip it and try to go to the next one (along the Eulerian cycle).

It is easy to prove that the last step works. Moreover, thanks to the triangle inequality, each skipping at Step 4 is in fact a shortcut; i.e., the length of the cycle does not increase. Hence it gives us a TSP tour no more than twice as long as the optimal one.

The Christofides algorithm follows a similar outline but combines the minimum spanning tree with a solution of another problem, minimum-weight perfect matching. This gives a TSP tour which is at most 1.5 times the optimal. The Christofides algorithm was one of the first approximation algorithms, and was in part responsible for drawing attention to approximation algorithms as a practical approach to intractable problems. As a matter of fact, the term "algorithm" was not commonly extended to approximation algorithms until later; the Christofides algorithm was initially referred to as the Christofides heuristic.

In the special case that distances between cities are all either one or two (and thus the triangle inequality is necessarily satisfied), there is a polynomial-time approximation algorithm that finds a tour of length at most 8/7 times the optimal tour length. However, it is a long-standing (since 1975) open problem to improve the Christofides approximation factor of 1.5 for general metric TSP to a smaller constant. It is known that, unless P = NP, there is no polynomial-time algorithm that finds a tour of length at most 220/219=1.00456... times the optimal tour's length. In the case of bounded metrics it is known that there is no polynomial time algorithm that constructs a tour of length at most 321/320 times the optimal tour's length, unless P = NP.

Euclidean TSP

The Euclidean TSP, or planar TSP, is the TSP with the distance being the ordinary Euclidean distance. The Euclidean TSP is a particular case of the metric TSP, since distances in a plane obey the triangle inequality.

Like the general TSP, the Euclidean TSP (and therefore the general metric TSP) is NP-complete. However, in some respects it seems to be easier than the general metric TSP. For example, the minimum spanning tree of the graph associated with an instance of the Euclidean TSP is a Euclidean minimum spanning tree, and so can be computed in expected O(n log n) time for n points (considerably less than the number of edges). This enables the simple 2-approximation algorithm for TSP with triangle inequality above to operate more quickly.

In general, for any c > 0, where d is the number of dimensions in the Euclidean space, there is a polynomial-time algorithm that finds a tour of length at most (1 + 1/c) times the optimal for geometric instances of TSP in time; this is called a polynomial-time approximation scheme (PTAS). Sanjeev Arora and Joseph S. B. Mitchell were awarded the Gödel Prize for their concurrent discovery of a PTAS for the Euclidean TSP in 2010. In practice, heuristics with weaker guarantees continue to be used.

Asymmetric TSP

In most cases, the distance between two nodes in the TSP network is the same in both directions. The case where the distance from A to B is not equal to the distance from B to A is called asymmetric TSP. A practical application of an asymmetric TSP is route optimisation using street-level routing (which is made asymmetric by one-way streets, slip-roads, motorways, etc.).

Solving by Conversion to Symmetric TSP

Solving an asymmetric TSP graph can be somewhat complex. The following is a 3×3 matrix containing all possible path weights between the nodes A, B and C. One option is to turn an asymmetric matrix of size N into a symmetric matrix of size 2N.

Table: *Asymmetric path Weights*

	A	*B*	*C*
A		1	2
B	6		3
C	5	4	

To double the size, each of the nodes in the graph is duplicated, creating a second ghost node. Using duplicate points with very low weights, such as ", provides a cheap route "linking" back to the real node and allowing symmetric evaluation to continue. The original 3×3 matrix shown above is visible in the bottom left and the inverse of the original in the top-right. Both copies of the matrix have had their diagonals replaced by the low-cost hop paths, represented by $-\infty$.

Table: *Symmetric path Weights*

	A	*B*	*C*	*A′*	*B′*	*C′*
A				$-\infty$	6	5
B				1	$-\infty$	4
C				2	3	$-\infty$
A′	$-\infty$	1	2			
B′	6	$-\infty$	3			
C′	5	4	$-\infty$			

The original 3×3 matrix would produce two Hamiltonian cycles (a path that visits every node once), namely A-B-C-A [score 9] and A-C-B-A [score 12]. Evaluating the 6×6 symmetric version of the same problem now produces many paths, including A-A2 -B-B2 -C-C2 -A, A-B2 -C-A2 -A, A-A2 -B-C2 -A [all score 9 $-\infty$].

The important thing about each new sequence is that there will be an alternation between dashed (A2,B2,C2) and undashed nodes (A, B, C) and that the link to "jump" between any related pair (A-A2) is effectively free. A version of the algorithm could use any weight for the A-A2 path, as long as that weight is lower than all other path weights present in the graph. As the path weight to "jump" must effectively be "free", the value zero (0) could be used to represent this cost—if zero is not being used for another purpose already (such as designating invalid paths). In the two examples above, non-existent paths between nodes are shown as a blank square.

Benchmarks

For benchmarking of TSP algorithms, TSPLIB is a library of sample instances of the TSP and related problems is maintained. Many of them are lists of actual cities and layouts of actual printed circuits.

Human Performance on TSP

The TSP, in particular the Euclidean variant of the problem, has attracted the attention of researchers in cognitive psychology. It is observed that humans are able to produce good quality solutions quickly. The first issue of the Journal of Problem Solving is devoted to the topic of human performance on TSP.

TSP path length for random pointset in a square: Suppose N points are randomly distributed in a 1 x 1 square with N>>1. Consider many such squares. Suppose we want to know the average of the shortest path length (i.e. TSP solution) of each square.

Lower Bound

$\frac{1}{2}\sqrt{N}$ is a lower bound obtained by assuming i be a point in the tour sequence and i has its nearest neighbour as its next in the path.

$\left(\frac{1}{4}+\frac{3}{8}\right)\sqrt{N}$ is a better lower bound obtained by assuming is next nearest, and is previous second nearest.

$\sqrt{\frac{N}{2}}$ is an even better lower bound obtained by dividing the path sequence into two parts as before_i and after_i with each part containing N/2 points, and then deleting the before_i part to form a diluted pointset.

- David S. Johnson obtained a lower bound by computer experiment: $0.7080\sqrt{N}+0.522$, where 0.522 comes from the points near square boundary which have fewer neighbours.
- Christine L. Valenzuela and Antonia J. Jones obtained another lower bound by computer experiment: $0.7078\sqrt{N}+0.551$,

Upper Bound

By applying Simulated Annealing method on samples of N=40000, computer analysis shows an upper bound of

$\left(\frac{\sqrt{N}}{2}+0.72\right)\cdot 1.015$, where 0.72 comes from the boundary effect.

Because the actual solution is only the shortest path, for the purposes of programmatic search another upper bound is the length of any previously discovered approximation.

Analyst's Travelling Salesman Problem

There is an analogous problem in geometric measure theory which asks the following: under what conditions may a subset E of Euclidean space be contained in a rectifiable curve (that is, when is there a continuous curve that visits every point in E)? This problem is known as the analyst's travelling salesman problem or the geometric travelling salesman problem.

8

Chromatin and Chromosomes

Chromatin is the combination of DNA, histone, and other proteins that makes up chromosomes. It is found inside the nuclear envelope of eukaryotic cells. It is divided between heterochromatin (condensed) and euchromatin (extended) forms. The functions of chromatin are to package DNA into a smaller volume to fit in the cell, to strengthen the DNA to allow mitosis and meiosis, and to control gene expression and DNA replication. Changes in chromatin structure are affected by chemical modifications of histone proteins, such as methylation and acetylation, and by other DNA-binding proteins.

Basic Structure

In simple terms, there are three levels of chromatin organization:

- DNA wraps around histone proteins forming nucleosomes - the "beads on a string" structure.
- A 30 nm condensed chromatin fibre consisting of nucleosome arrays in their most compact form.
- Higher-level DNA packaging into the metaphase chromosome.

These structures do not occur in prokaryotic cells. Examples of cells with more extreme packaging are spermatozoa and avian red blood cells. During spermiogenesis, the spermatid's chromatin is remodelled into a more spaced-packaged, widened, almost crystal-like structure. This process is associated with the cessation of transcription and involves nuclear protein exchange. The histones are mostly displaced, and replaced by protamines (small, arginine-rich proteins).

It should also be noted that, during mitosis, while most of the chromatin is tightly compacted, there are small regions that are not as tightly compacted. These regions often correspond to promoter regions of genes that were active in that cell type prior to entry into cromitosis. The lack of compaction of these regions is called bookmarking, which is an epigenetic mechanism believed to be important for transmitting to daughter cells the "memory" of which genes were active prior to entry into mitosis. This bookmarking mechanism is needed to help transmit this memory because transcription ceases during mitosis.

During Interphase

The structure of chromatin during interphase is optimised to allow easy access of transcription and DNA repair factors to the DNA while compacting the DNA into the nucleus. The structure varies depending on the access required to the DNA. Genes that require regular access by RNA polymerase require the looser structure provided by euchromatin.

Change in Structure

Chromatin undergoes various forms of change in its structure. Histone proteins, the foundation blocks of chromatin, are modified by various post-translational modification to alter DNA packing. Acetylation results in the loosening of chromatin and lends itself to replication and transcription. When certain residues are methylated they hold DNA together strongly and restrict access to various enzymes.

A recent study showed that there is a bivalent structure present in the chromatin: methylated lysine residues at location 4 and 27 on histone 3. It is thought that this may be involved in development; there is more methylation of lysine 27 in embryonic cells than in differentiated cells, whereas lysine 4 methylation positively regulates transcription by recruiting nucleosome remodeling enzymes and histone acetylases.

Nucleocytoplasmic Exchange

Between prokaryotes and eukaryotes resides in the absence or presence of the nuclear envelope. There is the main differences. This structure can be considered as a second not only by observing the continuities of the two elements at many points but also by studying these structures in different stages of cellular activity.

The nuclear envelope is a derivative of the endoplasmic reticulum is observed during mitosis. At telophase, cisternae of the endoplasmic reticulum collect around the chromosomes to reform the nuclear envelope. The nuclear envelop consists of two concernic membranes separated by a perinuclear cisterna 10 to 15 nm in width. These membranes have a basic unit structure similar to that of the plasma membrane. The nuclear envelope consists of flattened cisternae of the endoplasmic reticulum having ribosomes only on the outer surface. At certain points the nuclear envelope is interrupted by structures called *pores.*

Around the margins of these nuclear pores both membranes are in continuty. At their nuclear side the pores are generally aligned with channels of nucleoplasm stituated between more condensed lumps of chromatin which attach to the inner membrane. The number of poes varies from 40 to 145 per square micrometer in nuclei of various plants and animals.

At their highest density the pores are packed in hexagonal array with centre-to-centre pore distances of about 150. The nuclear pares are vary large, 50 to 80 nm in diameter. In nuclei of *Mammalia* it has been calculated that nuclear pores account for 5 to 15 per cent of the surface area of the nuclear membrane.

In amphibian oöcytes, certain plant cells, and protozoa the surface occupied by the pores may be as high as 20 to 36 per cent. Nuclear pores, however, are not wide-open channels, the electron microscope has revealed that they are occluded by an electron-dense materia. The pores are enclosed by circular structures called *annuli.*

The pores and annuli are together designated the *pore complex.* Negative staining techniques have demonstrated that the annuli are octagonal in shape and are about 60 nm in diameter. Each annulus consists of eight *granules* about 15 nm, which are present on both the nuclear and cytoplasmic surfaces and a less defined, amorphous *annular material* in the opening itself.

This material is digested b y trypsin and remains unaffected when exposed to ribonuclease and deoxyribonuclease. These findings suggest that the annular material is protein in nature. The pore complex is apparently a rather rigid structure present in a fixed number according to cell type. In certain physiologic stages, however, they may change in number.

For example, they are reduced in number in maturing erythroblasts and in spermatids, and this correlates with the low transcriptional activity of these cells. In the frog *Xenopus laevis,* oöcytes (which are very active in transcription) have 60 pores/μm^2 (and up to 30 million pore complexes per nucleus), whereas mature erythrocytes (inactive in transcription) have only about 3 pores/μm^2 (and a total of only 150 to 300 pores per nucleus.

The nuclear envelope has been isolated from a variety of cells and analysed biochemically. The main conclusion to be derived from these studies is that the composition of the nuclear membrane is very similar to that of the rough endoplasmic reticulum.

This emphasizes the relationship between both membrane systems, which can also be visualized by their physical continuity and by the presence of ribosomes attached to the outer nuclear membrane. The developing oöcytes and spermatocytes, as well as in certain embroynic and tumor cells, the so-called *annulated lamellae* may be found in the cytoplasm. These membranous structures appear as stacks of cisternae or flattened sacs with an intrasaccular space 10 to 20 nm wide. At regular intervals these sacs are traversed by pore complexes that are similar in morphology and structure to those found in the nuclear envelope. These are closely spaced and may occupy up to 50 per cent of the surface of the membrane.

The lamellae have been interpreted as budding off the nuclear envelope. More precise work has demonstrated that first there is a budding of the outer membrane of the nuclear envelope into flattened sacs; later the pore complexes appear. They are usually associated with rapidly proliferating cells. The material exchanged between nucleus and cytoplasm must traverse the nuclear pore complexes. This exchange is very selective and allows passage of only certain molecules, of either low or very high molecular weight.

The permeability of the nuclear membrane with respect to small ions is reflected in some of the electrochemical properties of this structure, which can be investigated with fine microelectrodes. When giant cells from the salivary gland of *Drosophila* are penetrated with a microelectrode, there is an abrupt change in potential at the plasma membrane (— 12 mV); then, as the microelectrode enters the nucleus, there is another drop in negative potential at the nuclear membrane (—13 mV). These results suggest that the nuclear envelope is a diffusion barrier for ions as small as K^+, Na^+, or Cl^-. On the other hand, very large structures such as ribosomal subunits, which are assembled in the nucleolus, are able to leave the nucleus.

All RNAs can exit freely; heterogeneous nuclear RNA (hnRNA, a large precursor of mRNA which contains intervening sequences is never found in the cytoplasm, and the mRNA sequences can traverse the pores only after extensive processing has occurred within the nucleus. Xenopus oöcytes are particularly useful for studying nucleocytoplasmic exchanges because these cells are very large (up to 1.2 mm diameter), and substances can be easily microinjected into the cytoplasm. One can then find out whether or not these substances entered the nucleus by manually isolating the giant oöcyte nucleus (also called the germinal vesicle), which measures up to 0.4 mm in diameter. Studies involving the microinjection of dextrans, colloidal gold particles, and labeled proteins sugest that the oöcyte nuclear membrane has functional pors with a radius of about 4.5 nm. It was found that small proteins (such as lysozyme, MW 14,000; and ovalbumin, MW 44,000), were able to enter the nucleus, that bovine serum albumin (MW 62,500) entered extremely slowly, and that larger proteins did not enter at all. This upper limit for diffusion applies only to proteins that are not normally nuclear proteins for, as we will see in the following section, nuclearspecific proteins have differently.

The cell nucleus contains a specific subset of proteins. For example, RNA polymerase, DNA polymerase, and histones are located specifically inside the nucleus. However, all these proteins are synthesized in the cytoplasm and must be subsequently transported into the nucleus. Here we will discuss experiments which show that nuclear proteins contain, within the structure of the nature protein, a signal that enables them to accumulate in the nucleus after microinjection into the cytoplasm.

This is observed because when labeled proteins, such as histones or a preparation of nucleoplasmic proteins, are microinjected into the cytoplasm of an oöcyte, they are not only able to enter the nucleus but also to *accumulate* in it.

When the nuclear proteins of normal oöcytes are examined by two dimensional gel electrophoresis, two types of proteins can be distinguished:

1. Nucleus specific proteins,
2. Proteins common to *both* nucleus and cytoplasm.

Actin is the most conspicuous member of the "common-to-both" proteins, and it is one of the major constituents of the oöcyte nucleoplasm.

A labeled preparation of *Xenopus* hucleoplasmic proteins (which contained actin and nuclear-specific proteins. After 24 hours the oöcytes were manually separated into nucleus and cytoplasm, and the proteins were analysed by two-dimensional gels. It was found that nucleus-specific proteins were greatly concentrated in the nucleus (100-fold in the case of N1), while microinjected actin was found both in nucleus and cytoplasm. This is as expected, since actin is normally present in both cell compartments. The nuclearspecific proteins here in their molecular structures some type of signal that enables them to accumulate selectively in the nucleus. This signal must be absent in proteins which do not accumulate in the nucleus, such as actin.

Some of the nuclear proteins are quite large (N1 has a MW of 120,000 daltons) but are still able to re-enter the nucleus rapidly, it therefore seems that when proteins are nucleus-specific, factors other than size determine nuclear pore permeability. Cells have the capacity to place specific proteins into the corresponding cell compartments, as occurs with the mitochondria or chloroplast proteins that are synthesized in the cytoplasm, or with proteins that are exported from the cell. The mechanism by which proteins are secreted into the endoplasmic reticulum is very different from the one used for the accumulation of proteins inside the nucleus. The "signal peptide" enables the secreted proteins to traverse the endoplasmic reticulum membrane only as they are being synthesized, and it is rapidly removed thereafter; the nuclear proteins, however, accumulate in the nucleus independently of protein synthesis, and this is not a transient property but is intrinsic to the structure of the mature protein.

Genetic Constituent

Cell have the genetic constituent *i.e.*, DNA, carrying information in a coded from from cell to cell and from organism to organism. Within cells, DNA is not free but is complexed with proteins in a structure called chromatin.

The domonstration that nuclei contain amount of DNA which is a landmark in cell biology because it suggested that DNA was the genetic material and also that genetic information is not lost during the differentiation of the various somatic tissues. This was achieved using cytophotometric methods. The DNA in nuclei was stained using the Feulgen reaction, and the amount of stain was estimated by microcytophotometry. Cytophotometry has produced accurate measurements in a large variety of cell types within the same species, showing that the DNA content is practically constant in the different tissues. This is certainly related to the number of chromosome sets (n), which is 2n (diploid) in most somatic cells. However, certain tissues, like liver, have some cells that become polyploid (4n or 8n), and their nuclei have a correspondingly higher DNA content.

The other example which is provided by the study of spermatogenesis. Before meiosis occurs, there are two classes of cells (spermatogonia) having different DNA contents (2n and 4n). The early primary spematocyte (meiotic prophase) has a tetraploid amount of DNA. After the first meiotic division, the secondary spermatocyte contains half the DNA content, corresponding to a diploid condition. Finally, the second meiotic division results in four spermatids; these have the DNA of only one chromosomal set (haploid cell). Similar results have been observed in oögenesis.

Each species has a characteristic amount of DNA. Eukaryotes vary greatly in DNA content but always contain much more DNA than prokaryots. Lower eukaryotes in general have less DNA, such as the nematode *C. elegans,* which has only 20 times more DNA than *Escherichia coli,* or the fruit fly *Drosophila melanogaster*, which has 40 times more. Vertebrates have greater DNA content, in general about 700 times more than *E. coli.*

The highest DNA content is that of the salamander *Amphiuma,* Which can contain 168 picograms of DNA in a single nucleus, *i.e.*, nearly 40,000 times more than *E. coli.* It is considered that one pictogram of DNA is equivalent to 31 cm of DNA, it is possible to calculate that there are about 174 cm of DNA in human diploid cells, 37 meters in *Trillium* and 97 meters in polytenic chromosomes of *Drosophila*. The DNA content in the 46 human chromosomes has been estimated by cytophotometry, and from these measurements it appears that the DNA content is proportional to the size of the chromosome. The largest chromosome, which is 10 mm long, should accommodate about 7.2 cm of DNA in a tightly packed form; *i.e.*, the DNA is compacted about 7000-fold in a metaphase chromosome! In this chapter we will analyse how this folding is achieved. During interphase, chromatin can be present in a condensed state (heterochromatin) or in a more dispersed form (euchromatin). The heterochromatic regions tend to form aggregates, sometimes called chromocentres.

Thin-section electron microscopy contributed little to our knowledge of the organization of chromatin in the interphase nucleus, in contrast to the results obtained by electron microscopic analysis of *spread* preparations. Asure have already shown in previous chapter, the distribution of chromatin in a nucleus stained with uranyl acetate. Both RNA and DNA are heavily stained, but DNA is most conspicuous.

The placement of the chromatin in clumps near the nuclear envelope and the interchromatin channels leading to the nuclear pores are clearly visible. The mass of the nucleolus is of a lesser density and is surrounded by an associated chromatin of a greater density; the nucleoplasm is light and reveals fine granules.

The nucleus has also been stained with uranyl acetate, but it was then treated with EDTA (ethylenedi-aminetetraacetic acid) which preferentially removes the uranyl ions from the DNA. EDTA is a chelating agent which binds to the phosphate groups of DNA, thus displacing the UO_2^{2+} ions. In this way, the RNA-containing structures of the nucleus and the ribosomes in the cytoplasm become more conspicuous. DNA structure shows a high electron density.

Chromatin can be isolated biochemically by purifying nuclei and then lysing them in hypotonic solutions. When prepared in this way, chromatin appears as a viscous, gelatinous substance which contains DNA, RNA, basic proteins called histones, and nonhistone (more acidic) proteins.

The content of RNA and nonhistone protein is variable between different chromatin preparations, but histone and DNA are always present in a fixed ratio of about one to one. The nonhistone proteins are very heterogeneous; they vary in different tissues and include RNA and DNA polymerases, among other enzymes, and putative regulatory proteins. On the other hand, there are only five types of histones, each one present in large amount and this suggested that they could have a structural role. Some of the properties of histones. Histones are small proteins which are basic because they have a high content (10 to 20 per cent) of the basic amino acids arginine and lysine. Being basic, histones bind tightly to DNA, which is acid. The four main histones, H2A, H2B, H3, and H4, are very similar in different species. For example, the sequence of histone H3 from the rat differs only in two-amino acids from that of peas. These four histones are present in equimolar amounts (two of each every 200 base pairs of DNA).

Histone H1 is not conserved between species and has tissue-specific forms; is present only once per 200 base pairs of DNA; is rather loosely associated with chromatin (if can be eluted from DNA by adding low concentrations of salt); and is not a component of the nucleosome but is involved in the maintenance of a higher-order folding of chromatin. Prokaryotes do not have histones, but a basic protein that might serve the same function has been isolated from *E. coli*. When eukaryotic nuclei were *spread* on EM specimen holders (instead of prepared as thin sections), it was found that chromatin has a repeating structure of beads about 10 nm in diameter connected by a string of DNA. Most, if not all, of the DNA is present in this form. The beads-on-a-string appearance is not the true structure of chromatin but is an artifact arising from the way in which the sample is prepared, which results in a loss of histone H1. With more mild treatments (which do not remove H1), the beads are not stretched but can be observed as a 10 nm fibre, with the beads touching each other. The 10nm fibre therefore represents the first level of organization of chromatin within cells.

The Nucleosome

Studies of digestion of chromatin by nucleases 22 showed that the DNA was cut into multiples of a unit size, which was later found to be about 200 base pairs of DNA in length; *i.e.*, the DNA was cut at intervals of 200, 600, 800 base pairs and so forth. Other studies showed that histones H3 and H4 tend to associate in solution forming tetramers consisting of two of each. Since the four histones are in equimolar amounts (two of each per 200 base pairs of DNA), in 1974 R. Kornberg (son of Arthur Kornberg, the discoverer of DNA polymerase) proposed a model for the nucleosome, in which the four histones H2A, H2B, H3, and H4 are arranged in octamers containing two of each of them, every 200 base pairs of DNA.

The DNA is cut into discrete fragments by micrococcal nuclease, because the linker regions between two neighboring nucleosomes are more exposed to the enzyme. Very extensive digestion leads to the degradation of the linker DNA and generates a nucleosome *core particle*, which contains 140 base pairs of DNA and lacks the 60 base pairs of the linker DNA. Core particles lack histone H1, indicating that H1 is normally associated with the linker DNA. Core nucleosomes have been crystallized, and x-ray diffraction and EM studies have shown that the nucleosome is a flat disk about 11 nm in diameter and 5.7 nm in height. The DNA is on the outside and turns twice around each histone octamer.

The length of DNA per nucleosome can vary in different tissues. For example, chicken erythrocytes have a repeat length of 210 base pairs, and chicken liver has about 195 base pairs per repeat. The reason for this is not clear, but it seems to be related to the type of histone H1 in each tissue. The variability is restricted to the length of the linker segment, and all tissues have a constant amount of DNA 140 base pairs) in the nucleosome core. When nucleosomes are in close apposition in the 10 nm filaments, the packing of DNA is about five-to seven-fold, *i.e.*, five to seven times more compact than free DNA.

However, this is still 1000-fold lower than the packing ratio of DNA in metaphase chromosomes, which is achieved by further folding the 10 nm fibre. Underlectron microscopic studies of chromatin fibres in interphase nuclei and mitotic chromosomes; have revealed a "*thick fibre*" of a diameter that varies from 20 to 30 nm and which probably represents the structure of inactive chromatin. You can seek that the 20 to 30 nm fibre consists of closely packed nucleosomes.

The 20 to 30 nm fibre probably arises from the folding of the nucleosome chain into a solenoidal structure having about six nucleosomes per turn. The histone H1 is located in the central "hole" of the solenoid, and the whole structure is stabilized by interactions between different H1 molecules. Solenoidal structures can be formed in vitro by adding Mg^{2+} to 10 nm fibre preparations, but they can not be formed when histone H1-depleted chromatin is treated in the same way.

The DNA of a 20 to 30 nm solenoid has a packing that is about 40-fold. However, the DNA of a metaphase chromosome is packed between 5000 and 10,000 times; *i.e.*, the 20 to 30 nm fibre must be further folded more than 100-fold. At present, we know very little of how this feat is achieved.

The Chromosomes

A chromosome is an organized structure of DNA and protein that is found in cells. It is a single piece of coiled DNA containing many genes, regulatory elements and other nucleotide sequences. Chromosomes also contain DNA-bound proteins, which serve to package the DNA and control its functions.

Chromosomes vary widely between different organisms. The DNA molecule may be circular or linear, and can be composed of 10,000 to 1,000,000,000 nucleotides in a long chain. Typically eukaryotic cells (cells with nuclei) have large linear chromosomes and prokaryotic cells (cells without defined nuclei) have smaller circular chromosomes, although there are many exceptions to this rule. Furthermore, cells may contain more than one type of chromosome; for example, mitochondria in most eukaryotes and chloroplasts in plants have their own small chromosomes.

In eukaryotes, nuclear chromosomes are packaged by proteins into a condensed structure called chromatin. This allows the very long DNA molecules to fit into the cell nucleus. The structure of chromosomes and chromatin varies through the cell cycle. Chromosomes are the essential unit for cellular division and must be replicated, divided, and passed successfully to their daughter cells so as to ensure the genetic diversity and survival of their progeny. Chromosomes may exist as either duplicated or unduplicated—unduplicated chromosomes are single linear strands, whereas duplicated chromosomes (copied during synthesis phase) contain two copies joined by a centromere.

Compaction of the duplicated chromosomes during mitosis and meiosis results in the classic four-arm structure (pictured to the right). Chromosomal recombination plays a vital role in genetic diversity. If these structures are manipulated incorrectly, through processes known as chromosomal instability and translocation, the cell may undergo mitotic catastrophe and die, or it may unexpectedly evade apoptosis leading to the progression of cancer.

In practice "chromosome" is a rather loosely defined term. In prokaryotes and viruses, the term genophore is more appropriate when no chromatin is present. However, a large body of work uses the term chromosome regardless of chromatin content. In prokaryotes DNA is usually arranged as a circle, which is tightly coiled in on itself, sometimes accompanied by one or more smaller, circular DNA molecules called plasmids. These small circular genomes are also found in mitochondria and chloroplasts, reflecting their bacterial origins. The simplest genophores are found in viruses: these DNA or RNA molecules are short linear or circular genophores that often lack structural proteins.

During cell division chromatin becomes condensed into the chromosomes. Since chromosomes can be clearly visualized with the optical microscope, they were the subject of much study soon after their discovery in 1876, and by 1910 it became clear that genetic phenomena could be explained in terms of chromosomes behaviour. For this very same reason, the study of chromosomes remains of the utmost importance even today. In this section we will analyse the general morphology and composition of chromosomes.

Chromosomes can be studied in tissue sections but they are best visualized in *squash* preparations. Fragments of tissues are stained with basic dyes (*e.g.*, orcein or Giemsa) and then squashed between slide and coverslip by gentle pressure. Sometimes hypotonic solutions are used prior to squashing to produce swelling of the nucleus and a better separation of the individual chromosomes. The morphology of the chromosomes is best studied during metaphase and anaphase, which are the periods of maximal contraction.

As per classification, there are four types according to their shape, which, in turn, is determined by the position of the centromere (point of attachment to the mitotic spindle). *Telocentric* chromosomes have the centromere located on one end; *acrocentric* chromosomes have a very small or even imperceptible short arm; *submetacentric* chromosomes have arms of unequal length; and *metacentric* chromosomes have equal or almost equal arms. During anaphase movements the chromosomes bend at the centromere, so that metacentric chromosomes are v-shaped, and acrocentric chromosomes are rod-shaped. This angular deviation can also be observed at metaphase (provided that the cells are not treated with colchicine).

As per classification, cytogeneticists produced many names to describe the various components that can be visualized in mitotic chromosomes. The student will probably find this terminology initially confusing, but it is important to become familiar with these names, for they not only define particular structures but also specific properties of the chromosomes.

Chromomeres

Chromomeres are bead like accumulations of chromatin material that are sometimes visible along the chromonema. Chromomeres are especially obvious in polytene chromosomes, where they become aligned side by side, constituting the chromosome bands. These tightly folded regions of the chromonema are of considerable interest, because they are thought to correspond to the units of genetic function in the chromosome. At metaphase the chromosome is tightly coiled, and the chromomeres are no longer visible.

Centromere

The centromere lies within a thinner segment of the chromosome, the *primary constriction*. The regions flanking the centromere frequently contain highly repetitive DNA and may stain more intensely with basic dyes (heterochromatin).

Kinetochore

The centromere surrounded by kinetcore shaped. Unlike the other chromosome components, the kinetochore can be studied best in thin sections with the electron microscope, which reveals a trilaminar structure. The kinetochore has an dense outer layer, middle layer of low density, and a dense inner layer tightly attached to the centromere.

The function of the kinetochore is to provide a centre of assembly for microtubules; when isolated metaphase chromosomes are incubated in a solution of tubulin with an appropriate buffer, microtubules are reassembled at the kinetochores of the chromosome.

Chromatid

At metaphase each chromosome consists of two symmetrical structures; the *chromatids;*, each one of which contains a single DNA molecule. The chromatids are attached to each other only by the centromere and become separated at the start of anaphase, when the sister chromatids migrate to opposite poles. Therefore, *anaphase chromosomes have only one chromatid, while metaphase chromosomes have two.*

Chromonemata

During prophase and some times during interphase the chromosomal material becomes visible as very thin filaments, which are called chromonemata and which represent chromatids in early stages of condensation. "Chromatid" and "chromonema," therefore, are two names for the same structure: a single linear DNA molecule with its associated proteins. The vast majority of chromosomes have only one centromere. Some species have diffuse centromeres, with microtubules attached along the length of the chromosome (*holocentric chromosomes*).

In some chromosomal abnormalities (induced for example by x-rays), chromosomes may break and fuse with other ones producing chromosomes without centromeres (*acentric*) or with two centromeres (*dicentric*). Both types of aberration are unstable; one because it cannot attach to the mitotic spindle and remains in the cytoplasm, the other because the two centromeres tend to migrate to opposite poles, thus leading to chromosomal fragmentation.

Telomere

The term 'Telomere applies to the tips of the chromosomes, which have distinct cytological properties. When chromosomes are broken by x-rays, the free ends without telomeres become "sticky" and fuse with other broken chromosomes; they will not, however, fuse with a normal telomere. Stable deletions in chromosomes are always interstitial and preserve the original telomere. In addition to promoting chromosomal stability, in some species telomeres mediate transient end-to-end association of chromosomes. A telomeric

DNA sequence which is present at the tips of all the *Drosophila melanogaster* chromosomes has been isolated. Telomeres contain the ends of the long linear DNA molecule contained in each chromatid.

Secondary Constrictions

These are the other morphologic characteristics. Constant in their position and extent, these constrictions are useful in identifying particular chromosomes in a set. Secondary constrictions are distinguished from the primary constriction by the absence of marked angular deviations of the chromosomal segments.

Nucleolar Organizers

These areas are certain secondary constrictions which contain the gene coding for 18S and 28S ribosomal. The and which induce the formation of nucleoli. The secondary constriction arises because the rRNA genes are transcribed arises because the rRNA genes are transcribed very actively, interfering with chromosomal condensation. In man, the nucleolar organizers are located in the secondary constrictions of chromosomes, 13, 14, 15, 21, and 22, all of which are acrocentrics and have satellites.

Satellites

Another morphologic element present in certain chromosomes is the satellite. This is a rounded body separated from the rest of the chromosomes by a secondary constriction. The satellite and the constriction are constant in shape and size for each particular chromosomes. The term Karyotype is concerned with the whole group of characteristics that allows the identification of a particular chromosomal set; *i.e.*, the number of chromosomes, relative size, position of the centromere, length of the arms, secondary constrictions, satellites, and so forth.

The karyotype is characteristic of an individual, species, genus, or larger grouping, and may be represented by a diagram in which the pairs of homologues are ordered in a series of decreasing size. Some species may have special characteristic; for example, the mouse has acrocentric chromosomes, many amphibia have only metacentric chromosomes, and plants frequently have heterochromatic regions at the telomeres.

Some species are particularly favorable for cytogenetic analysis: Salamanders and grasshoppers have high DNA content and very large chromosomes which produce very beautiful meiotic preparations; among plants, broad beans (*Vicia fava*) and onions (*Allium cepa*) have particularly large chromosomes which produce nice preparations of mitotic cells from root tips. Each chromatid represents a single linear DNA molecule with outs associated proteins. This concept, sometimes called the *unineme theory*, is supported by several lines of evidence semiconservative replication of a single DNA molecule. When cells are labeled with ^{3}H-thymidine or with bromode-oxyuridine (an analogue that becomes incorporated into DNA and prevents the staining of the chromatid with certain dyes) and then grown for two cycles in an unlabeled medium, only one of the granddaughter chromatids is labeled, as expected for a single DNA molecule.

Furthermore, exchanges between sister chromatids can be induced experimentally, and in this case every change in chromatid is accompanied by a reciprocal change in the other, a fact which is also consistent with the idea that there is only one DNA molecule with the finding that G_1 chromosomes (which have not yet replicated their DNA) have only one chromatid, and that G^2 chromosomes have two.

The unineme organization of the chromatid is supported directly by the isolation of DNA molecules that are long enough to contain all the DNA from a single chromatid. The length of these giant molecules can be estimated by viscoelastic measurements.

The two sister chromatids are mirror images one of the other. The morphologic characteristics of one chromatid, such as chromomeres, secondary constrictions, and satellites, always have their counterparts in the other. The reason for this symmetry is that sister chromatids contain identical DNA molecules, and the morphologic features of chromosomes are ultimately determined by the DNA sequence they contain.

The condensation and decondensation cycles of the chromatin fibres during mitosis and meiosis would be greatly faciliated if each chromosome were maintained in a definite space within the nuclear cavity during interphase, so as to avoid entanglements of the chromatin fibres.

Some evidence exists in plant cells to indicate that the chromosomes retain during interphase the same spatial order they had at telophase. This was shown to be true in onion chromosomes, which have characteristic heterochromatic regions, both at the centromeres and at the telomeres.

Throughout interphase the centromeres remain on one side of the nucleus, whereas the telomeres are on the opposite side, in the same orientation that occurs during telophase. It is even possible that chromosomes may be arranged in a fixed pattern with respect to each other as is suggested by the classical work of D. P. Costello, who analysed the first embryonic division in a turbellarian that has chromosomes that can be recognized by particular morphologic features. The chromosomes were arranged in a precise order during metaphase. The probability that such an arrangement would arise at random is less than 1 in 2.

Condensation and decondensation cycles. The interphase chromatin becomes condensed at mitosis, and the chromatids become increasingly shorter from prophase to metaphase, and they become decondensed at telophase. In some favorable organisms, such as the plants *Trillium* and *Tradescantia*, the optical microscope shows that mitotic condesation involves a helical coiling of the chromatids. In other species, including man, the coils become apparent only after certain pretreatments that partially open the tight helix.

Here is showing that mitotic chromosomes consist of chromatin fibres (20 to 30) nm in diameter. In these whole mount chromosomal preparations the chromosomes are dispersed on an air-water interface and then dehydrated by the critical-point drying method of Anderson. In this process the water content is replaced by liquid CO_2 in a pressure chamber, and then the specimen is heated above the *critical temperature* (*i.e.*, 31°C), at which liquid CO_2 is converted suddenly into gas.

The disorderly folding of the 20 to 30 nm fibres that can be seen in these preparations is an artifact due to the spreading on water. In the living cell, the chromatin fibres must be much more ordered, since an orderly structure can be packed into a much smaller space than a disorganized one. As discussed earlier, a metaphase chromosome is 5000 to 10,000 times shorter than the equivalent amount of free DNA. By comparison, the DNA in the head of bacteriophage T_4 is packed at a ratio of 520/1.

The DNA in mitotic chromosomes is contained in nucleosome chains. which are in turn coiled into 20 to 30 nm chromatin fibres, this represents a packing ratio of 40/1. The way in which the higher levels of packaging are achieved is not known, but some ideas about this are emerging from the study of histone-depleted chromosomes. Structural nonhistone proteins could be involved in the higher-order coiling of the 20 to 30 nm chromatin fibres. It is possible to remove the histones from metaphase chromosomes (by adding the polyanion dextran sulfate, which competes with the DNA for the basic histones).

Histone-depleted chromosomes have a central core or *scaffold*, surrounded by a halo made of loops of DNA. The scaffold is made of nonhistone proteins and retains the general shape of the metaphase chromosome. Each chromosome has two scaffolds, one for each chromatid, and they are connected together at the centromere region. When the histones are removed, the DNA, which was packed about 40-fold in the chromatin fibres, becomes extended and produces loops with an average length of 25 μm (75,000 base pairs). A in each loop the DNA exits from the scaffold and returns to an adjacent point. On the basis of these observations a model of chromosome structure has been proposed in which the DNA is arranged in loops anchored to the nonhistone scaffold. Since the lateral loops have 25 μm, of DNA, after contracting 40-fold in the 20 to 30 nm fibre, they would be only about 0.6 μm long, a length consistent with the diameter of metaphase chromosomes (1 μm) and with micrographs such as the one.

Other work, using very different methodology, has suggested that it is possible that even in interphase nuclei the chromatin might be organized in "domains" or loops containing about 80,000 base pairs of DNA. This raises the possibility that the DNA domains or loops might persist throughout the cell cycle. More work will be required before we understand the molecular organization of chromosomes.

Heterochro Matin Region

In 1928 Heitz defined *heterochromatin* as those regions of the chromosome that remain condensed during interphase and early prophase and form the so-called *euchromatin (Gr., eu,* true). Heitz followed cells throughout the cell cycle and found chromosomal segments that do not decondense.

The heterochromatic segments tend to show preterential localizations in the centromeres of most plants and animals, at the telomeres (especially in plants), or intercalated in the chromosomal arms (frequently adjacent to the nucleolar organizers). In other cases, whole chromosomes become heterochromatic.

The heterochromatic regions can be visulized in condensed chromosomes as regions that stain more strongly or weakly than the euchromatic regions, showing what is called a *positive* or a *negative heteropyknosis* of the chromosomes (Gr., *hetero +pyknosis,* different staining).It is thought that in heterochromatin the DNA remains tightly packed in the 20 to 30 nm fibre, which probably represents the configuration of transcriptionally inactive chromatin.

Heterochromatin

Two types of heterochromatin are generally recognized constitutive heterochromatin which is permanently condensed in all types, of cells, and *facultative* heterochromatin, which is condensed only in certain cell types or at special stages of development. Frequently in *facultative* heterochromatin one chromosome of the pair be comes either totally or partially heterochromatic. The best known case is that of the X chromosomes in the mammalian female, one of which is active and remains euchromatic, whereas the other is inactive and forms the sex chromatin, or...body, at interphase

Constitutive heterochromatin is the most common type of heterochromatin. Most chromosomes contain large blocks of heterochromatin at the centromeres, frequently comprising 5 to 10 per cent of the total chromosomal DNA and, in some extreme cases, such as in *Drosophila virilis,* it can constitute up to 40 per cent of the chromosome. This type of heterochromatin contains highly repeated DNA sequences, which might have a structural role in chromosomes.

A property common to all types of heterochromatin is *late replication.* When cells are given a brief pulse of ^{3}H-thymidine during the late S phase, the label is incorporated only into the heterochromatic segments, indicating that they replicate after the bulk of the DNA. It is generally agreed that condensed chromatin is inactive in RNA synthesis. For example, in the next chapter we will see that condensed mitotic chromosomes do not synthesize RNA. There is good genetic evidence indicating that genes contained in heterochromatic segments are not expressed. Here, we will examine three of these lines of evidence.

Some cats have a striking spotted coloring of the coat. Because of the peculiar coloured patches of the coat, these cats are known as *tortoiseshells.* Tortoiseshell cats are always female, never male. The reason for this become clear when Mary Lyon put forward the hypothesis that the patchy pigmentation is produced by a gene contained in the X chromosome, which becomes heterochromatic and inactive in some groups of cells put not in others.

One of the X chromosomes in mammalian females becomes inactivated early in embryogenesis on a random basis, so that adult animals are mosaics in which 50 per cent of the cells have the paternal X chromosome euchromatic and active, and the other 50 per cent have the maternal X chromosome in an active state. Another striking instance of facultative heterochromatin is found in the mealy bug *(planococcus citri),* an insect in which the males have a haploid set of chromosomes that are entirely heterochromatic. The heterochromatic set is the paternal one.

The genes in these paternal chromosomes are switched off, as demonstrated by the fact that it is not possible to induce lethal mutations in them, even by high does of radiation, whereas dominant lethals can be readily obtained by irradiation of the euchromatic set. In female mealy bugs both sets of chromosomes are euchromatic and both are genetically active; *i.e.*, lethal mutations can be induced in both the paternal and the maternal set. Genes can also be inactivated by translocation into heterochromatic regions of the chromosomes. This was first shown in *Drosophila melanogaster,* in which a gene coding for eye colour (w) can be inactivated by translocating it close to the centromeric heterochromatin.

The biological significance of the switching off of genes in condensed chromatin might be more universal than indicated in the limited examples discussed here. Multi-cellular organisms have many specialized tissues, and in each one of them a particular set of genes is active and others are inactive. For example, globin genes are expressed in red blood cells but are not expressed in other tissues. We do not know how entire sets of genes are switched off, but many believe that variable degrees of chromatin condensation could be involved in the maintenance of the differentiated state.

Constitutive heterochromatin is related to repetitive DNA. In many cells, it is possible to demonstrate the presence of special DNA fractions using gradients of cesium chloride. By using the satellite DNA as the template for in vitro synthesis, it is possible to obtain a ^{3}H-RNA can be hybridized to the satellite DNA not only in the test tube but also directly on a slide with the cells that contain this type of DNA.

The hybridization, also called *in situ hybridization,* was detected by autoradiography and localized specifically at the centromeric regions of all the chromosomes. This experiment demonstrates that the satellite DNA is present in the centromeric heterochromatin.

The Nucleolus

The presence of nucleoli, which appear as dense granules within the nucleus, was first described by Fontana in 1781. By the end of the 19 th century a relationship between the size of the nucleolus and the synthetic activity of the cell was postulated. It was found that nucleoli were small or absent in cells exhibiting little protein synthesis (sperm cells, blastomeres, muscle cells, and so forth), whereas they were large in oöcytes, neurons, and secretory cells — those in which protein synthesis is a prominent feature. In the living cell, nucleoli are highly refringent bodies.

This is the result of a large concentration of solid material, which may constitute 40 to 85 per cent of the dry mass. The light microscope generally reveals the nucleolus as structurally homogeneous, although small corpuscles or vacuoles are sometimes noted. The nucleolus stains with pyronine and other stains and absorbs ultraviolet light at 260 nm. Treatment with ribonuclease shows that the capacity to absorb this basophilic stain and ultraviolet ratiation depends on the presence of RNA.

The nucleolus may be surrounded by a ring of Feulgen-positive chromatin, which represents the heterochromatic regions of the chromosomes associated with the nucleolus. In large nucleoli some Feulgen-positive granules can be seen in portions of the chromosomes that penetrate the nucleolus. The nucleolus is formed at the nucleolar organizer, which is a chromosomal site that contains tandem repeats of the genes coding for 18S and 28S and rRNA. This DNA becomes uncoiled and penetrates the nucleolus, where it is actively transcribed.

In addition, other ribosomal components, such as 5S RNA and the ribosomal proteins, which are synthesized in other parts of the cell, converge on the nucleolus, where the assembly of ribosomal subunits starts. Electron microscopy has confirmed the existence of a definite submicroscopic organization within the nucleolus. In some cells nucleoli have a compact structure; in others the structure may be more or less open and include clear regions that communicate with the nucleoplasm. Two characteristic components may be recognized: the granular zone and the fibrillar zone.

Granular Zone

The *granular zone* consists of electron-dense granules of 15 to 20 nm, *i.e.*, smaller in diameter than the ribosomes. Frequently the granular zone occupies the more peripheral region of the nucleolus, which is surrounded by the associated chromatin. Both the granular and the fibrillar zones are most clearly distinguished. The *fibrillar zone* consists of fine fibres 5 to 10 nm in diameter. It generally occupies the central region of the nucleolus. Both the granular and fibrillar zones are digested by ribonuclease. In some cells the nucleolus is surrounded by a ring of heterochromatin, called the *nucleolarassociated* chromatin, which sometimes extends into the nucleolus. Autoradiographic studies indicate that ^{3}H-uridine is incorporated first into the fibrillar area and is only later found in the granular part. This suggests the relationship:

$$\text{nucleolar DNA} \rightarrow \text{fibrillar area} \rightarrow \text{granular area}$$

The fibrillar area contains the ribosomal DNA and its initial transcripts (called 45S RNA), and the granular part contains ribosomal precursor particles at different stages of maturation. During mitosis nucleoli undergo cyclic changes. Nucleoli are formed around the DNA loop that extends from the nucleolar organizer. There may be several nucleoli per cell, but frequently they tend to fuse into one or few nucleoli. During late prophase the DNA loop containing the rRNA genes gradually retracts and coils into the nucleolar organizer of the corresponding chromosome. Since this DNA is very extended as a consequence of intense RNA synthesis, the nucleolar organizer region is one of the last to undergo condensation, thus producing a *secondary constriction* on the chromosome. The fibrillar and granular components are gradually dispersed into the nucleoplasm. After the cell divides, during telophase, the nucleolar organizer DNA uncoils and the nucleolus is reassembled.

A knowledge of the behaviour of the chromosomes during meiosis and especially in what has already been said about the genetic consequences. However, in 1856 when Gregor johann Mendel discoverd the fudamental laws of inheritance, nothing was known about the chromosomes and meiosis. His discovery was based on the precise quantification of experimental crosses and exceptional abstract thinking. He studied crosses between peas (*Pisum sativum*) which had pairs of differential or contrasing characteristics. For example, he used plant that have white and red flowers, smooth and roguh seeds, yellow and green seeds, long and short stems, and so forth. After crossing the parental generation (P_1), he observed the resulting *hybrids* of the first filial generation, F1. Then he crossed the hybrids (F_1) among themselves an studied the result in the second filial generation. F_2. In a cross between parents with yellow and green seeds in the first generation Mendel found that all the hybrids had yellow seeds and thus, the characteristic of only one parent In the second cross (F_2), the characteristics of both parents reappeared in the proportion of 75 percent to 25 per cent, or 3:1. Mendel postulated that the colour of the seeds was controlled by a "factor" that was transmitted to the offspring by means of the gametes. This hereditary factor, which is now called the *gene,* could be transmitted without mixing with othr genes. He postulated that the gene could be segregated in the hybrid into different games to be *distributed* in the offspring of the hybrid.

For this reason this is called the *law* or *principle of segregation of the genes.* Mendel found that the plants with yellow colour, had different genetic constutions. One-third of this group always gave yellow seeds, but the other two-thirds o the F_2 generation produced planats with yellow and green seeds in the ratio of 3:1 When the 25 per cent of plants in F_2 with green seeds wrer a pure strain for this character. If we represent the genes in the crossing by letters, designating by *A* the gene with yellow character and by *a* the gene with green character, we have the folloing.

In the first generation (F_1) both *A* and *a* genes are present, but only A is revealed because it is *dominant;* gene *a* remains hidden and is called *recessive.* In the hybrid F_1 both genes are segregated and enter different gametes. Half of them will have the gene *A,* and the other half,*a.* Since each individual produces two types of games in each sex, there are four possible combinations in F_2. This gives as a result the proportion 1:2:1, corresponding to 25 per cent of plants with pure yellow seeds (*AA*), 50 per cent with hybrid yellow seed (*Aa*), and 25 per cent with pure green seeds (*aa*).

Mendel's results can now be explained in terms of the behavioural of chromosomes and genes. The genes present in the chromosomes are found in pairs called *alleles.* In each homologous chromosome the gene for each trait occurs at a particular point called a *locus* (plural *loci*). The gray mouse will have two GG genes, one in each homologue. Since the two homologues pair and then separate to enter the gametes. The mechanism is the same in a dominant as in a recessive white mouse. In the hybrid F_1, one chromosome bears gene G and the homologous chromosome bears gene *g*. When hybrids are self-fertilized, the gametes unite in the combinations shown by the checkerboard method illustrated in the individuals having two similar alleles are called *homozygous* (i. e., *GG* or *gg*); those with different alleles, *heterozygous*.

Genotype and Phenotype

Johanssen proposed the term *genotype* for the genetic constitution and *phenotype* for the visible characteristics shown by the individual in (91). For example, in the case of the peas with green or yellow seeds there are two phenotypes in $F_{2:}$ yellow seeds and green seeds in the ratio of 3:1, respectively. However, according to the genetic constitution, there are three different genotypes: *AA, Aa,* and *aa* in the ratio of 1:2:1.

The phenotype includes all characteristics of the individual that are an expression of gene activity. For example, in humans phenotypic characteristics include eye colour, baldness, the different hemoglobins, the blood groups, and a difference in the ability to taste thiourea. In crossings of certain plants that have white and red flowers, such as *Mirabilis jalapa,* it is possible to find in F_2 three phenotypes (red, pink, and white flowers), which correspond to the three genotypes. This is due to incomplete dominance. The principle of dominance and recessiveness in not always accomplished completely; dominance may be complete in most cases, but incomplete in others. In this case there is a mixture of characteristics, called *intermediary heredity.*

Whereas the law of segregation applies to the behaviour of a single pair of genes, the *principle of independent assortment* describes the simultaneous behaviour of two or more pairs of genes located in different pairs of chromosomes. Genes that lie in separate chromosomes are independently distributed during meiosis. The resulting off-spring is a hybrid (also called a dihybrid) at two loci.

All the above-mentioned examples of genetic crosses illustrated the fact that during meiosis there is a random distribution of the chromosomes which leads to the segregation of the genes in the gametes. However, when this type of study was carried out in the fruit fly *Drosophila melanogaster* by Morgan and collaborators, it became evident that the law of independent assortment was not universally applicable and that in certain crosses of two or more allelic pairs of genes, there was a certain limitation of free segregation. In each case there was a marked tendency for parental combinations to remain linked and for a lesser proportion of new combinations to be produced. In *Drosophila* there are only four pairs of chromosomes, and this increases the chances for genes to occupy loci in the same chromosome. If two genes A and B, with the corresponding alleles a and b, are in the same chromosome, only two classes of gametes will be obtained. After studying a considerable number of different crosses in *Drosophila,* Morgan reached the conclusion that all genes of this fly were clustered into four linked groups corresponding to the four pairs of chromosomes.

Recombination During Prophase

Further studies along these lines revealed that the linkage is not absolute and that it may be broken with a certain frequency. We have seen in the previous chapter that *genetic crossing over* or *recombination* takes place between the DNA molecules and is not visible microscopically. However, the *chiamata* observed at diplonema morphologically represent in some way the sites of these molecular events.

Thus, there should be a correlation between the two phenomena. It is also known that the frequency of recombination of two linked genes is a function of the distance which separates them along the chromosome. When two genes are close to one another, the probability of crossing over is less than when they are far apart. If the distance between genes is estimated by linkage analysis, it is possible to construct a map indicating the relative position of each gene along the chromosome. An appraisal of the possible number of new recombinations may be obtained by counting the number of chiasmata during meiosis. The so-called *recombination index* is calculated by adding to the number of bivalents the number of chiasmata detected in the same cell at diplonema. In a species with a higher indes, the possibility of new recombinations is higher, and this implies a greater possibility of variation.

There are several other pieces of evidence to support the relationship between chiasmata and crossing over. The number of chiasmata is related to the length of the chromosomes in the bivalent. The presence of one chiasma reduces the possibility of another occurring in the immediate vicinity. This phenomenon has been called *positive interference.* The chiasma frequency is roughly constant in a given species, but can be modified by genetic or environmental action. Chromosome pairing, chiasma formation, and crossing over are under genetic control. Lines of rye with high and low chiasma frequencies have been obtained by inbreeding. Among environmental factors that may affect the number of chiasmata are temperature, radiation, chemicals, and nutrition.

Neurospora

Among the different organisms studies in genetics, the mold *Neurospora* osccupies a special place.

The advantage of this material is twofold:

- It is possible to identify and to follow the fate of each of the four chromatids present in the bivalent meiotic chromosome and thus to determine whether the crossing over involves two, three, or all four chromatids.

- It is possible to make a close correlation between genetic constitution and biochemical expression of genes.

The four cells resulting from the two meiotic division undergo a mitotic division, which gives rise to eight haploid ascospores. Each of these ascospore can be isolated by dissection and cultured separately, giving rise to haploid individuals having the genetic constitution carried in each of the four original chromatids of the bivalent chromosome.

Cytogenetics

The normal functioning of the genetic system is maintained by the constancy of the hereditary material carried in the chromosomes. We mentioned on Chapter all have already studied that the species has a diploid number of chromosomes which are in homologous pair, and the normal human karyotype will be studied in detail. Under various conditions there may be changes in the karyotype, with different genetic consequences. Chromosomes may *change in number,* maintaining a normal structure, or they may have *sructural alterations,* frequently called *chromosomal aberrations.* Such structural changes may appear spontaneously, or they may be caused by the action of ionizing rediations or by chemcals. Sometimes these changes may have a genetic component. For example, in the hereditary disease *xeroderma pigmentosum,* chromosomes are extremely sensitive to ultraviolet rediation.

Euploidy

There are two main kinds of change in the number of chromosomes. In *euploids,* the change is in the haploid number, but the entire set of chromosomes is kept balanced, whereas in *aneuploids* there is a loss or gain of one or more chromosomes causing the set to become unbalanced. In humans aneuploidy may cause severe alterations of the phenotype.

Some exceptional plants and animals have a *monoploid* (or *haploid)* chromosomes set. In these organisms meiosis is irregular because of the absence of homologous chromosomes. As a result, gametes with varying numbers of chromosomes may be formed.

Polyloidy

A plant or animal that has more than two haploid sets of chromosomes is called a *polyplid.* This change is common in nature especially in the flowering plants. A diploid organism has two similar genomes; a triploid has three; an autoterapoid has for; and so on. Polyploids may originate either by reduplication of the chromosome number inn somatic tissue with suppression of cytokinesis or by formation of gametes with an unreduced number of chromosomes. Meiosis in a triploid is more irrgular than in a tetraploid. In general, polyploids of uneven number are sterile because the gametes have a more unbalanced number of chromosomes.

The scarcity of polyploids among animals is due the mechanism by which sex is determined. If polyploidy occurs, the genic balance between the sexchromosomes and the autosomes is disturbed, and the race or species may disappear because of sterility.

Several species of amphibians show spontaneous polyploidy. In such polyploids the AND content is correspondingly increased. Polyploids are useful in the study of the expression of genes in multiple dosage. Studies of this type have been made on amphibians for serum albumin, hemoglobin, and various enzymes. Polyploidy has been induced experimentally by temperature shock. it is also possible to induce polyploidy with substances such as colchicine. acenaphthene, heteroauxin, and veratrine.

These substances inhibit the formation of the spindle, and thus cell division is not completed. After a time the cells recover their normal activity, but have double the number of chromosomes. Form the standpoint of pure and applied scientific work innumerable possibilities are offered by the experimental production of polypoids.

Allopolyploidy

This is a type of chromosomal variation that is produced in crosses between two species having different sets of chromosomes. The resulting hybrid has a different number of chromosomes. The resulting hybrid has a deferents. For example, the Argentine black *Sorghum (S. almum)* is an allotetraploid (2n=4×=40) originated in nature by an interspecific cross between *S. sudanense* (2n=2×=20). In this case the fertilization occurred between one abnormal diploid gamete of *S sudanense* and a normal gamete of the other species. In most cases crosses between distantly related species produce sterile diploid bybrids. *Aneuploidy* is produced by a failure in the separation of chromosomes during meiosis, called *non-disjunction. Monosomic* individuals have lost one of the chromosomes of the karyotype, and in *trisomics* there is a gain of one chromosome. These changes are studied specifically in the following chapter on the human karyotype.

Aneuploidy may also occure in somatic cells inn which the non-disjunction occurs at mitosis various tissues or parts of the body that have different chromosome numbers, *e.g.*, in mosaic individuals, variegation, and gynandromorphs.

In cultures of normal cells chromosomes variations are common, particularly after several transplants. These changes may lead to malignancy, but this is not the case in all cultures. The cytogenetic analysis of mammalian tumours has led investigators to consider them as altered karyotypes, which are genetically and cytologically unstable.

To understand the mechanism by which chromosomal aberrations are produced the student should recall the folded fibre model of chromosomes structure and the concept of *mononemic* structure of chromatids. According to this concept, each chromatid is made of one molecule of DNA; thus in the G_1 period there is a single DNA duplication that each chromosome contains two chromatids and two DNA molecules. In a chromatids and each chromosomal aberration there is an alteration in the structural organization of the chromosome.

For example, a break in the unit fibre may lead to a loss of material, called a *deletion,* and in cases inn which there are two breaks, a *translocation,* an *inversion,* or an *exchanges between sister chromatids* may be produced. In all these cases the chromosomes change can be observed microscopically, especially by observing the chromosomes at the stages of condensation. Chromosomal aberrations should be differentiated from *gene mutations* (also called *point mutations*), in which the change are produced at the molecular level. In these, the alteration resides in the genetic code (*i.e.*, in the sequence of DNA bases) and can be detected only by its genetic or molecular expression.

DNA

Both gene muations and chromosomal aberrations may occur spontaneously, but their frequency increases by the action of so-called ionizing rediation (*i.e.*, x-rays, B-reys, y-reys and ultraviolet light, as well as accelerated particles such as fast neutrons, protons, and soforth). (Other *mutagenic agents* are chemical substances, viruses, mycoplasms, and temperature changes. In various organisms it has been demonstrated that the number of mutations induced by radiation is proportional to the dose. In the case of *Drosophila,* the relation is a linear function in the range 25 to 9000 roentgen units (R).

The effects of radiation are cumulative over long periods of time. For example, 0.1 R per day for 10 years is enough to increase the mutation rate to about 150 per cent of the spontaneous level. In contrast to gene mutations, chromosomal aberrations increase exponentially. More chromosomal aberrations are produced by continuous than by intermittent treatment. A low dose of radiation may not be enough to cause fractures in one chromosome. As the dose increases, the number of breaks increases, and "aberrant" fusions becomes more and more likely, If the dose is intermittent or of low intermittent or of low intensity, there is a greater chance that the broken ends of the chromosome will rejoin, or "heal," in the original chromosome structure before a second break could cause aberration.

The type of chromosomal aberration depends on the period of the cell cycle the cell was in at the time of irradiation, A *chromatid break* occurs in the G_2 period, and *chromosomal break* (*i.e.*, two chromatids) occurs if the cell was in the G_1 period, which is before DNA duplication. Irradiation, generally produces localized lesions which may stabilize and from the so-called gaps. After some time, such gaps may be repaired. Such repair may result in complete restitution of the original structure (true repair). If there is no repair, the lesion becomes stabilized and cytologically visible. The process of restitution is inhibited by cold, cyanide, and dinitrophenol. Oxygen increases the number of fractures and chromosomal interchanges. During breakage of chromosome, the two fragments may either reunite or remain separated permanently. If, instead of reconstituting at once, the broken ends reduplicate to from two chromatids, the fragments with the centromere migrate to different poles, and the fragments without a centromere (acentric) are eliminated in the cytoplasm. Sometimes the sister fragements reduplicate and unite, forming a dicentric chromosome and another acentric chromosome. During mitosis, the centromeres of the dicentric chromosome move to opposite poles and form a "bridge" between the daughter nuclei that finally break at some point.

In man, double of mutiple chromosomal breaks are induced by acute exposures (*e.g.*, heavy medical irradiations, nuclear accidents, or nuclear warfare), whereas single breaks are produced at low doses. Chromosomal aberrations have been observed inn blood cultures of humans who have had rediation treatments or injections of radioactive substances. The genetic effect of radiation has been studied in space flights. After Geminals III and IV. no increase in chromsomal aberrations was found in the blood cells of the astronauts who made the flights

Mutations in *somatic cells* are not transmitted from generation to generation but may be cumulative and produce severe changes in the individual, depending on the type of cell affected and the time at which the mutation occurs. Radiation can affect tissues that undergo mitosis, as well as tissues in which cell division no longer takes place. It mutation occurs during early embryonic development, a large number of cells are affected. The majority of mutated genes are recessive and thus no effect as long as the individual is heterozygous.

It is probable that some cases of cancer produced in irradiated individuals are caused by somatic mutation. In contrast to somatic mutation, mutations inn the germ cells may be transmitted to the offspring. However, in most cases of mutation caused by irradiation, both somatic and germ cells are frequently affected. Even the lowest doses of rediation are genetically harmful, and the effect are dangerous to all organisms, from the simplest to man. In microorganisma and plants, however, a few useful mutations may be obtained by irradiation. For example, irradiation is one method of producing new antibiotics and plants with high economic value.

Deficiency

Deletion or loss of chromosomal material may be either *terminal or interstitial,* originating from one or two breaks, respectively, produced in the G_1 period. Some interstitial deletions may originate both a ring chromosome and an acentric rod, which is eliminated.

In heterozygous deficiency, one chromosome is normal, but its homologue is deficient. Animals with a homozygous deficiency usually do not survive to an adult stage because a complete set of genes is lacking. This suggests that most genes are indispensable at least in a single dose to the development of a viable organism Deficiencies are important in cytogenetic investigations of gene location of for determination of the presence and position of unmatched genes.

Duplication

Duplication occurs when a segment of the chromosome is represented two or more times. The duplicated fragment may be free, with a centromere of its own, or it may be incorporated into a chromosomal segment of the normal complement, If the fragment includes the centromere, it may be incorporated as a small extra chromosome. In general, duplication are less deleterious to the individual than deficiencies.

Translocation

In the so-called *reciprocal translocation,* segment are exchanged between non-homologous chromosomes. Such translocation may be homozygotic or heterozygotic. Cytologically, a translocated homozygote cannot be distinguished from a normal pair of chromosome, but it can be detected by genetic experiments. Heterozygotic translocations give rise to special pairing configurations in meiosis. An interesting result occurs when during translocation, both chromosome are broken vary close to their centromeres. The fusion creates a metanentric chromosome with two arms in the form of a V. and a small fragment, which tends to be eliminated.

Centric fusion has occurred during the phylogeny of *Drosophila,* grasshoppers, reptiles, birds, mammals, and other groups. It is a process that establishes a new type of chromosome and reduces the somatic chromosome number of the species. In the following chapter we will mention the importance of reciprocal translocations in human cytogenetics.

Inversion

An inversion is a chromosomal aberration in which a segment is inverted 180 degrees. Inversions are called pericenteric when the segment includes the centromere, and paracentric if the centromere is located outside the segment. In these aberrations there is typical configuration at pachynema consisting of a loop that allows the pairing of the inverted segement. In the paracentric inversion acentric and dicentric chromatids are formed.

Isochromosomes

A new type of chromosome may arise from a break (i. e., misdivision) at the centromere. The two resultant telocentric chromosomes may open up to produce chromosomes with two identical arms (i. e., isochromosome). This type of chromosome has been produced in irradiated material. At meiosis they may pair with themselves or with a normal homologue.

A *sister chromatid exchange* is an interchange of DNA between sister chromatids in chromosome, presumably involving DNA breakage followed by fusion. Sister chromatid exchanges are difficult to find using common cytologic methods because the chromatids are morphologically identical.

Such chromatid exchanges were first described in studies in which ^{3}H-thymidine was added during a replicating cycle which was followed by another cycle in a non-radioactive medium. As was mentioned in earlier this method demonstrated that the duplication of DNA is semiconservative however, in a few mitoses observed by radioautography, an alteration of the labelling along the chromatids was found and this was interpreted as the result of chromatid exchange caused by the radioactivity of the tritiated label used.

Analysis of this phenomenon has been greatly facilitated by the of *bromodeoxyuridine* (Brd U), a thymidine analogue that can be incorporated into the DNA of replicating cells instead of the original base. If BrdU is followed by a fluorescent day, the fluorescence of the segments that contain BrdU is greatly diminished in comparison with those with the original base. Furthermore, there is also similarly decreased staining with the Giemsa stain.

The use of this technique, however, has been unable to discover whether the chromatid exchange could occur spontaneously or whether it is induced by the BrdU. It has, though, been of great help in differentiating the various inherited diseases characterized by chromosomal fragility, which have an increased frequency of sister chromatid exchanges and a tendency to have associated neoplasia. Some of these diseases (*e.g.*, Bloom's syndrome, Fanconis anemia, and ataxia-telangiectasia) are presumably related to defects in DNA repair.

Sister chromatid exchange has also been important inn studying the effect of mutagens on the chromosomes. Various mutagenic drugs that are alkylating agents, such as *mitomycin* C and nitrogen mustered, produce a great number of breaks and chromatid exchages. The intimate association of sister chromatid exchange with mutagenesis and carcinogenesis may have important medical implications.

Fundamental Role of Chromosome in Evolution

The development of comparative cytology and cytogenetics has brought about great progress towards an understanding of evolution. MuClung and S. Navashin were the first to emphasize the importance of cytogenetics to toxnomy and to the study of evolution by comparing genomes of related species. Systematics has been greatly advanced by cytogenetic investigation, which now provides many of the best methods for elucidating correlations between different taxonomic categories. In general, families, genera, and species are characterized by different genetic systems.

The study of the karyotypes of different species has revealed interesting facts about both the plant and animal kingdoms. It has been demonstrated that individuals in wild populations are, to some extent, heterogeneous cytologically and genetically. In some cases, even if the genes are identical they may be ordered in a different way, owing to alteration of the chromosomeal segments. These changes have ann important bearing on the evolution of species.

The majority of plant species originate from an abrupt and rapid change in nature, and aneuploidy and polyploidy are the prime sources of variation. In the animal kingdom polyploidy is not so important. Among vertebrates, different species of fishes have different number of chromosomes. Amphibians are generally characterized by a specific number for each family. Reptiles and bird have large chromosomes (macrochromosomes) and small chromosomes (microchr-omosomes) that serve to differentiate them cytologically.

Matthey distinguishes between the basic chromosome number and the number of chromosomal arms, also called the fundamental number (FN). According to this concept, the metacentric chromosome has *two* arms and acrocentric and metacentric chromosomes, and the number of arms in each the different species can be compared.

Another method used to study the cytogenetics of evolution is the application of measurements of total chromosomal area and DNA content. With regard to the absolute size of their chromosomes, mammals and birds constitute two independent groups. These two orders have different DNA contents and different sexdetermining mechanisms. Speciation depends more of chromosomal rearrangement and mutation of individual genes than on changes in the total amount of genetic content. Two opposite changes in the number (and configuration) of chromosomes are of particular importance in evolution. In *centric fusoin,* aprocess that leads to decrease in chromosomes join together to produce a metacentric chomosome. In *dissociation, or fission,* a process that leads to an increase in chromosome number, metacentric (commonly large) and a small supernumerary metacentric fragment become translocated, so that two acrocentric or submetacentric chromosomes are produced.

Fusion and dissociation are the main mechanisms by which the chromosome number can be decreased and increased during evolution of the majority of animals and in some groups of plants. In recent years the karyotype of the great apes, chimpanzee, gorilla, and oragutan, have been compared with that of man. It was found that these primates have 48 chromosomes. and attempts were made to correlate the chromosome fairs with those of the 46 human chromosomes. This analysis has been reatly facilitated by the use of banding techniques. Using the quinacrine technique it was demonstrated that human chromosomes 1, 3, 5, 6, 7, 8, 10, 11, 12, 13, 14, 15, 19, 20, 21, 22, X and Y had direct counterparts in the chimpansee, but has less similarity with the paris of the gorills and orangutan. The difference between the 48 pair karyotype and the 46 is due to the fusion of two acrocentric chromosomes, present in the hominoid apes, to make the human chromosome 2[7].

At present, pericentric inversion appears to be the main structural difference between individual chromosomes of the great apes and man. It has been postulated that evolution of the lruman karyotype has occurred by series of percentric inversions which permitted genetic isolation of small breeding groups and selection of favourable gene combinations that gave rest to *Home sapiens.* The homologies in morphology of the chromosomes are now being correlated with the mapping of genes in similar chromosome of man and the great apes. Studies of somatic and polytene chromosomes in several hundred species of *Drosophila* have elucidated the formulation and evolution of this elution of this genus, which has been thoroughly analysed from, genetic, ecologic, and geographic standpoints. Observation of chromosomal organization and of the different karyotypes in the individual, the species genera, and the major systematic groups indicates that a chromosomal mechanism is involved in the process of evolution.

The problem of evolution, however, is very complex and should be considered from the different biochemical, cytologic, genetic, ecologic, and experimental aspects. All these methods and approaches should be used to analyse the intricate relationships between groups off organisms, particularly those that show marked variations.

Functions of Chromosomes

The discovery that DNA is the prime genetic molecule, carrying all the hereditary information within chromosomes, immediately focused attention on its structure. It was hoped that knowledge of the structure would reveal how DNA carries the genetic messages that are replicated when chromosomes divide to produce two identical copies of themselves. During the late 1940s and early 1950s, several research groups in the United States and in Europe engaged in serious efforts—both cooperative and rival—to understand how the atoms of DNA are linked together by covalent bonds and how the resulting molecules are arranged in three-dimensional space. Not surprisingly, there initially were fears that DNA might have very complicated and perhaps bizarre structures that differed radically from one gene to another.

Great relief, if not general elation, was thus expressed when the fundamental DNA structure was found to be the double helix. It told us that all genes have roughly the same three-dimensional form and that the differences between two genes reside in the order and number of their four nucleotide building blocks along the complementary strands.

Now, some 50 years after the discovery of the double helix, this simple description of the genetic material remains true and has not had to be appreciably altered to accommodate new findings. Nevertheless, we have come to realise that the structure of DNA is not quite as uniform as was first thought. For example, the chromosome of some small viruses have single-stranded, not double-stranded, molecules. Moreover, the precise orientation of the base pairs varies slightly from base pair to base pair in a manner that is influenced by the local DNA sequence.

Some DNA sequences even permit the double helix to twist in the left-handed sense, as opposed to the right-handed sense originally formulated for DNA's general structure. And while some DNA molecules are linear, others are circular.

Still additional complexity comes from the supercoiling (further twisting) of the double helix, often around cores of DNA-binding proteins. Likewise, we now realise that RNA, which at first glance appears to be very similar to DNA, has its own distinctive structural features. It is principally found as a single-stranded molecule.

Yet by means of intra-strand base pairing, RNA exhibits extensive double-helical character and is capable of folding into a wealth of diverse tertiary structures. These structures are full of surprises, such as non-classical base pairs, base-backbone interactions, and knot-like configurations.

Most remarkable of all, and of profound evolutionary significance, some RNA molecules are enzymes that carry out reactions that are at the core of information transfer from nucleic acid to protein. Clearly, the structures of DNA and RNA are richer and more intricate than was at first appreciated. Indeed, there is no one generic structure for DNA and RNA.

Polynucleotide Chains

The most important feature of DNA is that it is usually composed of two polynucleotide chains twisted around each other in the form of a double helix. The upper part presents the structure of the double helix shown in a schematic form. Note that if inverted 180°, the double helix looks superficially the same, due to the complementary nature of the two DNA strands.

The space-filling model of the double helix, shows the components of the DNA molecule and their relative positions in the helical structure. The backbone of each strand of the helix is composed of alternating sugar and phosphate residues; the bases project inward but are accessible through the major and minor grooves.

Let us begin by considering the nature of the nucleotide, the fundamental building block of DNA. The nucleotide consists of a phosphate joined to a sugar, known as 2'-deoxyribose, to which a base is attached. The sugar is called *2'*-deoxyribose because there is no hydroxyl at position 2' (just two hydrogens).

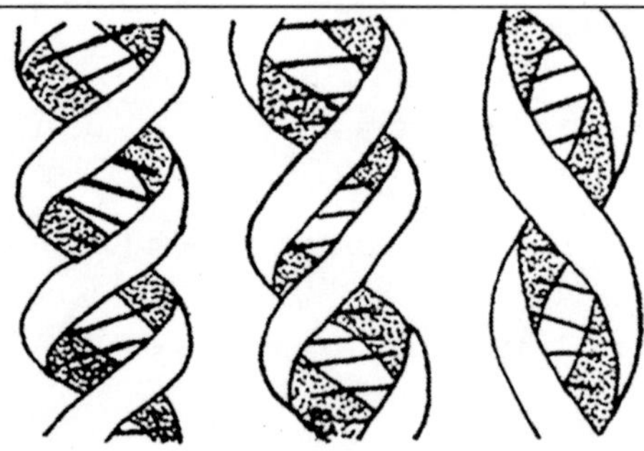

From	A	B	Z
Belical Twist	Right	Right	Left
Base Pairs Per Turn	11	10	12
Occurrence	RNA, DNA	DNA	DNA

Figure: *Polynucleotide Chains*

Note that the positions on the ribose are designated with primes to distinguish them from positions on the bases. We can think of how the base is joined to 2'-deoxyribose by imagining the removal of a molecule of water between the hydroxyl on the 1' carbon of the sugar and the base to form a glycosidic bond. The sugar and base alone are called a nucleoside. Likewise, we can imagine linking the phosphate to 2'-deoxyribose by removing a water molecule from between the phosphate and the hydroxyl on the 5' carbon to make a 5' phosphomonoester. Adding a phosphate (or more than one phosphate) to a nucleoside creates a nucleo*t*ide. Thus, by making a glycosidic bond between the base and the sugar, and by making a phosphoester bond between the sugar and the phosphoric acid, we have created a nucleotide.

Nucleotides are, in turn, joined to each other in polynucleotide chains through the 3' hydroxyl of 2'-deoxyribose of one nucleotide and the phosphate attached to the 5' hydroxyl of another nucleotide. This is a phosphodiester linkage in which the phosphoryl group between the two nucleotides has one sugar esterified to it through a 3' hydroxyl and a second sugar esterified to it through a 5' hydroxyl. Phosphodiester linkages create the repeating, sugar-phosphate backbone of the polynucleotide chain, which is a regular feature of DNA. In contrast, the order of the bases along the polynucleotide chain is irregular. This irregularity as well as the long length is, as we shall see, the basis for the enormous information content of DNA. The phosphodiester linkages impart an inherent polarity to the DNA chain.

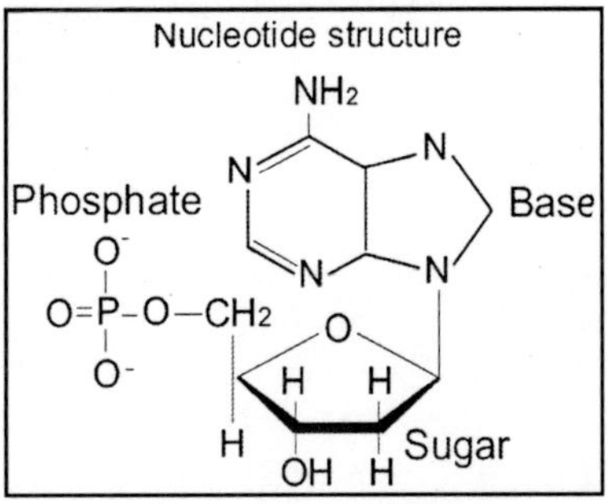

Figure: *Nucleotide*

This polarity is defined by the asymmetry of the nucleotides and the way they are joined. DNA chains have a free 5' phosphate or 5' hydroxyl at one end and a free 3' phosphate or 3' hydroxyl at the other end. The convention is to write DNA sequences from the 5' end (on the left) to the 3' end, generally with a 5' phosphate and a 3' hydroxyl.

Purines and Pyrimidines

The bases in DNA fall into two classes, purines and pyrimidines. The purines are adenine and guanine, and the pyrimidines are cytosine and thymine. The purines are derived from the double-ringed structure. Adenine and guanine share this essential structure but with different groups attached. The bases are attached to the deoxyribose by glycosidic linkages at N1 of the pyrimidines or at N9 of the purines.

Each of the bases exists in two alternative tautomeric states, which are in equilibrium with each other. The equilibrium lies far to the side of the conventional structures, which are the predominant states and the ones important for base pairing. The nitrogen atoms attached to the purine and pyrimidine rings are in the amino form in the predominant state and only rarely assume the imino configuration. Likewise, the oxygen atoms attached to the guanine and thymine normally have the keto form and only rarely take on the enol configuration. As we shall see, the capacity to form an alternative tautomer is a frequent source of errors during DNA synthesis.

Double Helix

The double helix is composed of two polynucleotide chains that are held together by weak, non-covalent bonds between pairs of bases. Adenine on one chain is always paired with thymine on the other chain and, likewise, guanine is always paired with cytosine. The two strands have the same helical geometry but base pairing holds them together with the opposite polarity.

That is, the base at the 5' end of one strand is paired with the base at the 3' end of the other strand. The strands are said to have an anti-parallel orientation. This anti-parallel orientation is a stereochemical consequence of the way that adenine and thymine and guanine and cytosine pair with each together.

Pairing of Adenine and Thymine

The pairing between adenine and thymine and between guanine and cytosine results in a complementary relationship between the sequence of bases on the two intertwined chains and gives DNA its self-encoding character. For example, if we have the sequence 5'-ATGTC-3' on one chain, the opposite chain must have the complementary sequence 3'-TACAG-5'. The strictness of the rules for this "Watson-Crick" pairing derives from the complementarity both of shape and of hydrogen bonding properties between adenine and thymine and between guanine and cytosine. Adenine and thymine match up so that a hydrogen bond can form between the exocyclic amino group at C6 on adenine and the carbonyl at C4 in thymine; and likewise, a hydrogen bond can form between N1 of adenine and N3 of thymine.

A corresponding arrangement can be drawn between a guanine and a cytosine, so that there is both hydrogen bonding and shape complementarity in this base pair as well. A G:C base pair has three hydrogen bonds, because the exocyclic NH_2 at C_2 on guanine lies opposite to, and can hydrogen bond with, a carbonyl at C_2 on cytosine. Likewise, a hydrogen bond can form between N_1 of guanine and N_3 of cytosine and between the carbonyl at C_6 of guanine and the exocyclic NH_2 at C_4 of cytosine. Watson-Crick base pairing requires that the bases are in their preferred tautomeric states. An important feature of the double helix is that the two base pairs have exactly the same geometry; having an A:T base pair or a G:C base pair between the two sugars does not perturb the arrangement of the sugars. Neither does T:A or C:G. In other words, there is an approximately twofold axis of symmetry that relates the two sugars and all four base pairs can be accommodated within the same arrangement without any distortion of the overall structure of the DNA.

Bonding of the Double Helix

The hydrogen bonds between complementary bases are a fundamental feature of the double helix, contributing to the thermodynamic stability of the helix and providing the information content and specificity of base pairing. Hydrogen bonding might not at first glance appear to contribute importantly to the stability of DNA for the following reason. An organic molecule in aqueous solution has all of its hydrogen bonding properties satisfied by water molecules that come on and off very rapidly. As a result, for every hydrogen bond that is made when a base pair forms, a hydrogen bond with water is broken that was there before the base pair formed. Thus, the net energetic contribution of hydrogen bonds to the stability of the double helix would appear to be modest. However, when polynucleotide strands are separate, water molecules are lined up on the bases. When strands come together in the double helix, the water molecules are displaced from the bases. This creates disorder and increases entropy, thereby stabilizing the double helix. Hydrogen bonds are not the only force that stabilizes the double helix.

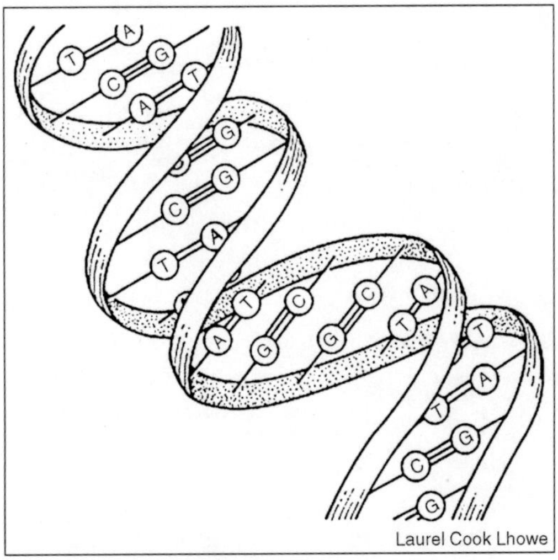

Figure: *Hydrogen Bonds Between Double Helix*

A second important contribution comes from stacking interactions between the bases. The bases are flat, relatively waterinsoluble molecules, and they tend to stack above each other roughly perpendicular to the direction of the helical axis. Electron cloud interactions (ð–ð) between bases in the helical stacks contribute significantly to the stability of the double helix. Hydrogen bonding is also important for the specificity of base pairing. Suppose we tried to pair an adenine with a cytosine.

Then we would have a hydrogen bond acceptor (N_1 of adenine) lying opposite a hydrogen bond acceptor (N_3 of cytosine) with no room to put a water molecule in between to satisfy the two acceptors. Likewise, two hydrogen bond donors, the NH_2 groups at C_6 of adenine and C_4 of cytosine, would lie opposite each other. Thus, an A:C base pair would be unstable because water would have to be stripped off the donor and acceptor groups without restoring the hydrogen bond formed within the base pair.

Left-Handed Helix

DNA containing alternative purine and pyrimidine residues can fold into left-handed as well as right-handed helices. To understand how DNA can form a left-handed helix, we need to consider the glycosidic bond that connects the base to the 1' position of 2'-deoxyribose. This bond can be in one of two conformations called *syn* and *anti*.

In right-handed DNA, the glycosidic bond is always in the *anti* conformation. In the left-handed helix, the fundamental repeating unit usually is a purine-pyrimidine dinucleotide, with the glycosidic bond in the *anti* conformation at pyrimidine residues and in the *syn* conformation at purine residues. It is this *syn* conformation at the purine nucleotides that is responsible for the left-handedness of the helix.

The change to the *syn* position in the purine residues to alternating *anti–syn* conformations gives the backbone of left-handed DNA a zigzag look, which distinguishes it from right-handed forms. The rotation that effects the change from *anti* to *syn* also causes the ribose group to undergo a change in its pucker. In solution alternating purine–pyrimidine residues assume the left-handed conformation only in the presence of high concentrations of positively charged ions (*e.g.*, Na^+) that shield the negatively charged phosphate groups. At lower salt concentrations, they form typical right-handed conformations. The physiological significance of Z DNA is uncertain and left-handed helices probably account at most for only a small of proportion of a cell's DNA.

DNA Strands

Because the two strands of the double helix are held together by relatively weak (non-covalent) forces, we might expect that the two strands could come apart easily. Indeed, the original structure for the double helix suggested that DNA replication would occur in just this manner. The complementary strands of double helix can also be made to come apart when a solution of DNA is heated above physiological temperatures (to near 100°C) or under conditions of high pH, a process known as denaturation. However, this complete separation of DNA strands by denaturation is reversible. When heated solutions of denatured DNA are slowly cooled, single strands often meet their complementary strands and reform regular double helices. The capacity to renature denatured DNA molecules permits artificial hybrid DNA molecules to be formed by slowly cooling mixtures of denatured DNA from two different sources.

Likewise, hybrids can be formed between complementary strands of DNA and RNA. The ability to form hybrids between two single-stranded nucleic acids (hybridization) is the basis for several indispensable techniques in molecular biology, such as Southern blot hybridization and DNA microarrays.

Important insights into the properties of the double helix were obtained from classic experiments carried out in the 1950s in which the denaturation of DNA was studied under a variety of conditions. In these experiments DNA denaturation was monitored by measuring the absorbance of ultraviolet light passed through a solution of DNA.

DNA maximally absorbs ultraviolet light at a wavelength of about 260 nm. It is the bases that are principally responsible for this absorption. When the temperature of a solution of DNA is raised to near the boiling point of water, the optical density (absorbance) at 260 nm markedly increases. The explanation for this increase is that duplex DNA is hypochromic; it absorbs less ultraviolet light by about 40 per cent than do individual DNA chains. The hypochromicity is due to base stacking, which diminishes the capacity of the bases in duplex DNA to absorb ultraviolet light. If we plot the optical density of DNA as a function of temperature, we observe that the increase in absorption occurs abruptly over a relatively narrow temperature range. The midpoint of this transition is the melting point or Tm.

Like ice, DNA melts: it undergoes a transition from a highly ordered double-helical structure to a much less ordered structure of individual strands. The sharpness of the increase in absorbance at the melting temperature tells us that the denaturation and renaturation of complementary DNA strands is a highly cooperative, zippering-like process.

Renaturation, for example, probably occurs by means of a slow nucleation process in which a relatively small stretch of bases on one strand find and pair with their complement on the complementary strand. The remainder of the two strands then rapidly zipper-up from the nucleation site to reform an extended double helix. The melting temperature of DNA is a characteristic of each DNA that is largely determined by the G:C content of the DNA and the ionic strength of the solution.

The higher the per cent of G:C base pairs in the DNA (and hence the lower the content of A:T base pairs), the higher the melting point. Likewise, the higher the salt concentration of the solution the greater the temperature at which the DNA denatures.

How do we explain this behaviour? G:C base pairs contribute more to the stability of DNA than do A:T base pairs because of the greater number of hydrogen bonds for the former (three in a G:C base pair versus two for A:T) but also importantly because the stacking interactions of G:C base pairs with adjacent base pairs are more favorable than the corresponding interactions of A:T base pairs with their neighboring base pairs.

The effect of ionic strength reflects another fundamental feature of the double helix. The backbones of the two DNA strands contain phosphoryl groups, which carry a negative charge.

These negative charges are close enough across the two strands that if not shielded they tend to cause the strands to repel each other, facilitating their separation. At high ionic strength, the negative charges are shielded by cations, thereby stabilizing the helix. Conversely, at low ionic strength the unshielded negative charges render the helix less stable.

DNA Molecules

It was initially believed that all DNA molecules are linear and have two free ends. Indeed, the chromosomes of eukaryotic cells each contain a single (extremely long) DNA molecule. But now we know that some DNAs are circles. For example, the chromosome of the small monkey DNA virus SV40 is a circular, double-helical DNA molecule of about 5,000 base pairs.

Also, most (but not all) bacterial chromosomes are circular; *E. coli* has a circular chromosome of about 5 million base pairs. Additionally, many bacteria have small autonomously replicating genetic elements known as plasmids, which are generally circular DNA molecules.

Interestingly, some DNA molecules are sometimes linear and sometimes circular. The most well-known example is that of the bacteriophage ‘, a DNA virus of *E. coli*. The phage ‘genome is a linear double-stranded molecule in the virion particle. However, when the ‘genome is injected into an *E. coli* cell during infection, the DNA circularizes. This occurs by base-pairing between single-stranded regions that protrude from the ends of the DNA and that have complementary sequences (“sticky ends”).

DNA Topology

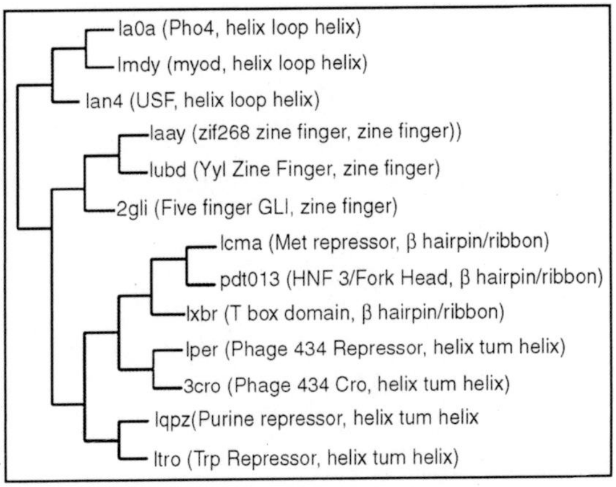

Figure: *DNA-Proteins*

As DNA is a flexible structure, its exact molecular parameters are a function of both the surrounding ionic environment and the nature of the DNA-binding proteins with which it is complexed. Because their ends are free, linear DNA molecules can freely rotate to accommodate changes in the number of times the two chains of the double helix twist about each other.

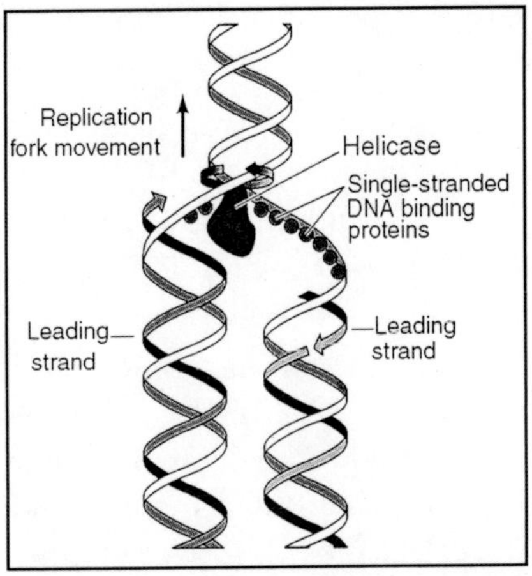

Figure: *DNA-binding Proteins*

But if the two ends are covalently linked to form a circular DNA molecule and if there are no interruptions in the sugar phosphate backbones of the two strands, then the absolute number of times the chains can twist about each other cannot change. Such a covalently closed, circular DNA is said to be topologically constrained. Even the linear DNA molecules of eukaryotic chromosomes are subject to topological constraints due to their entrainment in chromatin and interaction with other cellular components. Despite these constraints, DNA participates in numerous dynamic processes in the cell. For example, the two strands of the double helix, which are twisted around each other, must rapidly separate in order for DNA to be duplicated and to be transcribed into RNA.

Thus, understanding the topology of DNA and how the cell both accommodates and exploits topological constraints during DNA replication, transcription, and other chromosomal transactions is of fundamental importance in molecular biology.

Circular DNA

Let us consider the topological properties of covalently closed, circular DNA, which is referred to as cccDNA. Because there are no interruptions in either polynucleotide chain, the two strands of cccDNA cannot be separated from each other without the breaking of a covalent bond.

If we wished to separate the two circular strands without permanently breaking any bonds in the sugar phosphate backbones,we would have to pass one strand through the other strand repeatedly.

The number of times one strand would have to be passed through the other strand in order for the two strands to be entirely separated from each other is called the linking number. The linking number, which is always an integer, is an invariant topological property of cccDNA, no matter how much the DNA molecule is distorted.

The linking number is the sum of two geometric components called the twist and the writhe. Let us consider twist first. Twist is simply the number of helical turns of one strand about the other, that is, the number of times one strand completely wraps around the other strand. Consider a cccDNA that is lying flat on a plane.

In this flat conformation, the linking number is fully composed of twist. Indeed, the twist can be easily determined by counting the number of times the two strands cross each other. The helical crossovers (twist) in a right-handed helix are defined as positive such that the linking number of DNA will have a positive value. But cccDNA is generally not lying flat on a plane. Rather, it is usually torsionally stressed such that the long axis of the double helix crosses over itself, often repeatedly, in three-dimensional space. This is called *writhe.* To visualize the distortions caused by torsional stress, think of the coiling of a telephone cord that has been overtwisted. Writhe can take two forms. One form is the interwound or plectonemic writhe, in which the long axis is twisted around itself.

The other form of writhe is a toroid or spiral in which the long axis is wound in a cylindrical manner, as often occurs when DNA wraps around protein. The writhing number (*Wr*) is the total number of interwound and/or spiral writhes in cccDNA. Interwound writhe and spiral writhe are topologically equivalent to each other and are readily interconvertible geometric properties of cccDNA. Also, twist and writhe are interconvertible. A molecule of cccDNA can readily undergo distortions that convert some of its twist to writhe or some of its writhe to twist without the breakage of any covalent bonds. The only constraint is that the sum of the twist number (*Tw*) and the writhing number (*Wr*) must remain equal to the linking number (*Lk*). This constraint is described by the equation: $Lk = Tw + Wr$.

Linking of Fully Relaxed cccDNA

Consider cccDNA that is free of supercoiling (that is, it is said to be relaxed) and whose twist corresponds to that of the B form of DNA in solution under physiological conditions (about 10.5 base pairs per turn of the helix). The linking number (*Lk*) of such cccDNA under physiological conditions is assigned the symbol Lk^o. Lk^o for such a molecule is the number of base pairs divided by 10.5.

For a cccDNA of 10,500 base pairs, $Lk = +1{,}000$. (The sign is positive because the twists of DNA are right-handed.) One way to see this is to imagine pulling one strand of the 10,500 base pair cccDNA out into a flat circle. If we did this, then the other strand would cross the flat circular strand 1,000 times. How can we remove supercoils from cccDNA if it is not already relaxed? One procedure is to treat the DNA mildly with the enzyme DNase I, so as to break on average one phosphodiester bond (or a small number of bonds) in each DNA molecule. Once the DNA has been "nicked" in this manner, it is no longer topologically constrained and the strands can rotate freely, allowing writhe to dissipate. If the nick is then repaired, the resulting cccDNA molecules will be relaxed and will have on average an *Lk* that is equal to Lk^o. (Due to rotational fluctuation at the time the nick is repaired, some of the resulting cccDNAs will have an *Lk* that is somewhat greater than Lk^o and others will have an *Lk* that is somewhat lower. Thus, the relaxation procedure will generate a narrow spectrum of topoisomers whose average *Lk* is equal to Lk^o).

Supercoiling in Eukaryotes

DNA in the nucleus of eukaryotic cells is packaged in small particles known as nucleosomes in which the double helix is wrapped almost two times around the outside circumference of a protein core. We will be able to recognize this wrapping as the toroid or spiral form of writhe. Importantly, it occurs in a lefthanded manner. It turns out that writhe in the form of left-handed spirals is equivalent to negative supercoils. Thus, the packaging of DNA into nucleosomes introduces negative superhelical density.

Topoisomerases

As we have seen, the linking number is an invariant property of DNA that is topologically constrained. It can only be changed by introducing interruptions into the sugar-phosphate backbone.

A remarkable class of enzymes known as topoisomerases are able to do just that by introducing transient nicks or breaks into the DNA. Topoisomerases are of two broad types. Type II topoisomerases change the linking number in steps of two. They make transient double-stranded breaks in the DNA, through which they pass a region of uncut duplex DNA before resealing the break.

Type II topoisomerases require energy from ATP hydrolysis for their action. Type I topoisomerases, in contrast, change the linking number of DNA in steps of one. They make transient singlestranded breaks in the DNA, allowing one strand to pass through the break in the other before resealing the nick. Type I topoisomerases relax DNA by removing supercoils.

They can be compared to the protocol of introducing nicks into cccDNA with DNase and then repairing the nicks, which as we saw can be used to relax cccDNA, except that type I topoisomerases relax DNA in a controlled and concerted manner.

In contrast to type II topoisomerases, type I topoisomerases do not require ATP. Both type I and type II topoisomerases work through an intermediate in which the enzyme is covalently attached to one end of the broken DNA.

Both prokaryotes and eukaryotes have type I and type II topoisomerases, which are capable of removing supercoils from DNA. In addition, however, prokaryotes have a special type II topoisomerase known as DNA gyrase that is able to introduce negative supercoils, rather than remove them. DNA gyrase is responsible for the negative supercoiling of chromosomes in prokaryotes, which facilitates unwinding of the DNA duplex during transcription and DNA replication.

Electrophoresis

Covalently closed, circular DNA molecules of the same length but of different linking numbers are called DNA topoisomers. Even though topoisomers have the same molecular weight, they can be separated from each other by electrophoresis through a gel of agarose.

The basis for this separation is that the greater the writhe the more compact the shape of a cccDNA. Once again, think of how supercoiling a telephone cord causes it to become more compact. The more compact the DNA, the more easily (up to a point) it is able to migrate through the gel matrix. Thus, a fully relaxed cccDNA migrates more slowly than a highly supercoiled topoisomer of the same circular DNA. Molecules in adjacent rungs of the ladder differ from each other by a linking number difference of just one. Obviously, electrophoretic mobility is highly sensitive to the topological state of DNA.

Helical Periodicity

The observation that DNA topoisomers can be separated from each other electrophoretically is the basis for a simple experiment that proves that DNA has a helical periodicity of about 10.5 base pairs per turn in solution. Consider three cccDNAs of sizes 3990, 3995, and 4011 base pairs that were relaxed to completion by treatment with topoisomerase I. When subjected to electrophoresis through agarose, the 3990- and 4011-base-pair DNAs exhibit essentially identical mobilities.

Due to thermal fluctuation, topoisomerase treatment actually generates a narrow spectrum of topoisomers, but for simplicity let us consider the mobility of only the most abundant topoisomer (that corresponding to the cccDNA in its most relaxed state).

The mobilities of the most abundant topoisomers for the 3990- and 4011-base-pair DNAs are indistinguishable because the 21-base-pair difference between them is negligible compared to the sizes of the rings. The most abundant topoisomer for the 3995-base-pair ring, however, is found to migrate slightly more rapidly than the other two rings even though it is only 5 base pairs larger than the 3990-base-pair ring.

How are we to explain this anomaly? The 3990- and 4011- base-pair rings in their most relaxed states are expected to have linking numbers equal to Lk^o, that is, 380 in the case of the 3990-base-pair ring (dividing the size by 10.5 base pairs) and 382 in the case of the 4011-base-pair ring.

Because Lk is equal to Lk^o, the linking difference ('$Lk = Lk - Lk^o$) in both cases is zero and there is no writhe. But because the linking number must be an integer, the most relaxed state for the 3995-base-pair ring would be either of two topoisomers having linking numbers of 380 or 381. However, Lk^o for the 3995-base-pair ring is 380.5.

Thus, even in its most relaxed state, a covalently closed circle of 3995 base pairs would necessarily have about half a unit of writhe (its linking difference would be 0.5), and hence it would migrate more rapidly than the 3990- and 4011-base-pair circles. In other words, to explain how rings that differ in length by 21 base pairs (two turns of the helix) have the same mobility whereas a ring that differs in length by only 5 base pairs (about half a helical turn) exhibits a different mobility, we must conclude that DNA in solution has a helical periodicity of about 10.5 base pairs per turn.

Ethidium Ions Cause DNA to Unwind

Ethidium is a large, flat, multi-ringed cation. Its planar shape enables ethidium to slip (intercalate) between the stacked base pairs of DNA. Because it fluoresces when exposed to ultraviolet light, and because its fluorescence increases dramatically after intercalation, ethidium is used as a stain to visualize DNA.

When an ethidium ion intercalates between two base pairs, it causes the DNA to unwind by 26°, reducing the normal rotation per base pair from 36° to 10°. In other words, ethidium decreases the twist of DNA.

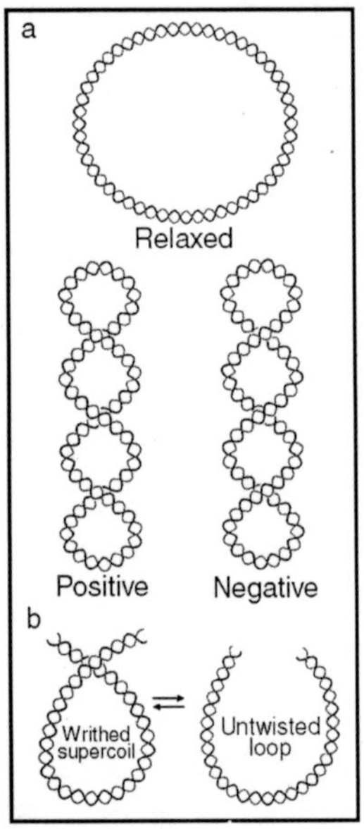

Figure: *Supercoiled DNA*

Imagine the extreme case of a DNA molecule that has an ethidium ion between every base pair. Instead of 10 base pairs per turn it would have 36! When ethidium binds to linear DNA or to a nicked circle, it simply causes the helical pitch to increase.

But consider what happens when ethidium binds to covalently closed, circular DNA. The linking number of the cccDNA does not change (no covalent bonds are broken and resealed), but the twist decreases by 26° for each molecule of ethidium that has bound to the DNA.

Because $Lk = Tw + Wr$, this decrease in Tw must be compensated for by a corresponding increase in Wr. If the circular DNA is initially negatively supercoiled (as is normally the case for circular DNAs isolated from cells), then the addition of ethidium will increase Wr.

In other words, the addition of ethidium will relax the DNA. If enough ethidium is added, the negative supercoiling will be brought to zero, and if even more ethidium is added, *Wr* will increase above zero and the DNA will become positively supercoiled.

Because the binding of ethidium increases *Wr*, its presence greatly affects the migration of cccDNA during gel electrophoresis. In the presence of non-saturating amounts of ethidium, negatively supercoiled circular DNAs are more relaxed and migrate more slowly, whereas relaxed cccDNAs become positively supercoiled and migrate more rapidly.

9

RNA Sequencing

While methods for each type of probing differ in steps, the probing of secondary structure involves certain steps in order to determine the structure. The first step involved is submitting the structural RNA to the probe of interest and incubating over a certain amount of time to allow the reaction to occur.

The RNA is then transcribed using reverse transcriptase PCR, where this results in different lengths of bands due to the modification of the RNA at specific sites, which causes the reverse transcriptase to fall off. These bands are then run on a gel with sequencing data determined from RNA sequencing.

Chemical Probing

RNA chemical probing can involve many different chemicals which serve to modify specific bases at certain sites to show certain locations available for specific modification type.

DMS

Dimethyl sulfate, known as DMS, is a chemical that can be used to modify nucleic acids in order to determine secondary structure. DMS modifies certain bases through methylation. One set of methylation products that is used for RNA is the methylation of N1 adenosine and N3 of cytosine which prevents the natural hydrogen bonds to form between bases modified by DMS.

This enables modification sites to be detected by rt-PCR, as modified sites cannot be basepaired, which results in the reverse transcriptase to fall off and different band sizes. DMS can only modify bases that are not basepaired, as single stranded nucleotides have the bases exposed so the chemical can modify. Detection of these modifications is done by examination of a gel and looking at bands present. One thing of note is that since modification prevents the addition of nucleotides at the modification site each of the bands generated is shifted by one base down on the gel.

The PCR products indicate which adenine and cytosine nucleotides are double stranded, that is protected from DMS modification, or single stranded, that is accessible to DMS modification. This can be used to begin to develop a model of secondary structure for the RNA molecule. Structure probing by DMS allows for detection of secondary structure changes due to binding of RNA molecules along with detecting changes in tertiary structure, but it lacks the ability to determine the origin of these changes, as well as protection of the RNA backbone.

DMS modification can also be used for DNA, for example in footprinting DNA-protein interactions.

CMCT

1-cyclohexyl-(2-morpholinoethyl)carbodiimide metho-p-toluene sulfonate known as CMCT is another chemical that is used in structure probing. CMCT like DMS serves to modify the exposed bases of specific nucleotides, which are uridine, and to a smaller extent guanine. CMCT reacts primarily with N3 of uridine and N1 of guanine modifying two sites responsible for hydrogen bonding on the bases.

Modification using CMCT is analogous in detection to DMS as modification prevents basepairing at specific nucleotides which are then detected using rt-PCR and running of the PCR products on an agarose gel. Structure probing of RNA by CMCT indicates the presence of uridine and guanine in single stranded regions by accessibility to modification or the presence of uridine and guanine in double stranded regions by protection from CMCT, which is the absence of a band. CMCT, like DMS can detect secondary and tertiary structure changes, but still has the same weaknesses as the method of modification is the same as DMS.

Kethoxal

The modification by kethoxal prevents single stranded guanine nucleotides to become modified, so that when reverse transcriptase reaches the modified guanine it falls off resulting in a band. After rt-PCR the products will be of different lengths where a band present indicates modification of a guanine base, and the absence of a band indicates that the base was not available for modification. Using this gel in combination with DMS and CMCT a model for structural RNAs can be formed through comparison between the gels which indicate protected and accessible positions in all the bases, which can be used to form a preliminary model.

SHAPE

Selective 22 -hydroxyl acylation analysed by primer extension, or SHAPE, takes advantage of reagents that preferentially modify the backbone of RNA in structurally flexible regions.

Reagents such as N-methylisotoic anhydride (NMIA) and 1-methyl-7-nitroisatoic anhydride (1M7) undergo hydrolysis to form adducts on the 2'-hydroxyl of the RNA backbone. Compared to the chemicals used in other RNA probing techniques, these reagents have the advantage of being largely unbiased to base identity, while remaining very sensitive to conformational dynamics.

Nucleotides which are constrained (usually by base-pairing) show less adduct formation than nucleotides which are unpaired. Adduct formation is quantified for each nucleotide in a given RNA by extension of a complementary DNA primer with reverse transcriptase and comparison of the resulting fragments with those from an unmodified control. SHAPE therefore reports on RNA structure at the individual nucleotide level.

This data can be used as input to generate highly accurate secondary structure models. SHAPE has been used to analyse diverse RNA structures, including that of an entire HIV-1 genome.

Hydroxyl Radical Probing

Probing with hydroxyl radicals involves an additional step, as hydroxyl radicals are short lived in solution they need to be generated. This can be done using H_2O_2, ascorbic acid, and Fe(II)-EDTA complex which is attached to the backbone through EDTA. The Fe(II) along with ascorbic acid generates hydroxyl radicals which then can react with the nucleic acid molecules. Hydroxyl radical probing is often used in conjunction with chemical probing of nucleic acid molecules that are thought to associate with proteins, this is due to the modification done by hydroxyl radicals. Hydroxyl radicals attack the ribose/deoxyribose ring and this results in breaking of the phosphate backbone, which is independent of secondary structure, as all backbone is accessible, but is instead resultant of protein or tertiary structure protection.

Probing with hydroxyl radicals shows the protection of structured nucleic acids by the proteins thought to be associated or folding on itself, where cleavage again results in a band formed through gel electrophoresis (after rt-PCR in the case of RNA) that is shorter than the full nucleic acid depending upon where it is cleaved. In this case since the last nucleotide is not modified the band length is indicative of the base that was cleaved. When examining the gel produced by running the gels on a band the areas of various strength of protection where areas of stronger protection for hydroxyl radicals can be said to have tighter association with a protein, or if no protein associates with the nucleic acid it can be caused by the tertiary fold.

In-line Probing

In-line probing does not involve treatment with any type of chemicals or reagents to modify exposed RNA structures. This type of probing assay uses the structure dependent cleavage of RNA, as areas that are single stranded are more flexible and over time will degrade, as RNA structure is not always stable. The process of in-line probing is often used to determine changes in structure due to ligand binding as this can result in different cleavage patterns.

The process of in-line probing involves incubation of structural or functional RNAs over a long period of time, can be several days, but varies in each experiment, and then running the incubated products on a gel to visualize the bands. This experiment is often done using two different conditions: 1) with ligand and 2) in the absence of ligand.

Cleavage results in shorter band lengths and is indicative of areas that are not basepaired, as those that are tend to be less sensitive to spontaneous cleavage. In line probing is a functional assay in ligand binding changes of RNA because it can show directly the change in flexibility and binding of regions of DNA in response to a ligand, as well as compare that response to analogous ligands. An in-line probing assay is then commonly used in dynamic studies, specifically when examining riboswitches.

Nucleotide Analog Interference Mapping

Nucleotide analog interference mapping (NAIM) is the process of using nucleotide analogs, molecules that are similar in some ways to nucleotides but lack function, to determine the importance of a functional group at each location of an RNA molecule. The process of NAIM is to insert a single nucleotide analog into a unique site.

This can be done by transcribing a short RNA using T7 RNA polymerase, then synthesizing a short oligonucleotide containing the analog in a specific position, then ligating them together on the DNA template using a ligase. The nucleotide analogs are tagged with a phosphorothioate, the active members of the RNA population are then distinguished from the inactive members, the inactive members then have the phophorothioate tag removed and the analog sites are identified using gel electrophoresis and autoradiography.

This indicates a functionally important nucleotides, as cleavage of the phosphorothioate by iodine results in an RNA that is cleaved at the site of the nucleotide analog insert. By running these truncated RNA molecules on a gel, the nucleotide of interest can be identified against a sequencing experiment Site directed incorporation results indicate positions of importance where when running on a gel, functional RNAs that have the analog incorporated at that position will have a band present, but if the analog results in non-functionality, when the functional RNA molecules are run on a gel there will be no band corresponding to that position on the gel.

This process can be used to evaluate an entire area, where analogs are placed in site specific locations, differing by a single nucleotide, then when functional RNAs are isolated and run on a gel, all areas where bands are produced indicate non-essential nucleotides, but areas where bands are absent from the functional RNA indicate that inserting a nucleotide analog in that position caused the RNA molecule to become non-functional.

Paracrystalline Lattice Models of B-DNA Structures

A paracrystalline lattice, or paracrystal, is a molecular or atomic lattice with significant amounts (e.g., larger than a few percent) of partial disordering of molecular arranegements. Limiting cases of the paracrystal model are nanostructures, such as glasses, liquids, etc., that may possess only local ordering and no global order. Liquid crystals also have paracrystalline rather than crystalline structures.

Highly hydrated B-DNA occurs naturally in living cells in such a paracrystalline state, which is a dynamic one in spite of the relatively rigid DNA double-helix stabilized by parallel hydrogen bonds between the nucleotide base-pairs in the two complementary, helical DNA chains.

For simplicity most DNA molecular models omit both water and ions dynamically bound to B-DNA, and are thus less useful for understanding the dynamic Behaviours of B-DNA *in vivo*. The physical and mathematical analysis of X-ray and spectroscopic data for paracrystalline B-DNA is therefore much more complicated than that of crystalline, A-DNA X-ray diffraction patterns. The paracrystal model is also important for DNA technological applications such as DNA nanotechnology. Novel techniques that combine X-ray diffraction of DNA with X-ray microscopy in hydrated living cells are now also being developed.

Genomic and Biotechnology Applications of DNA Molecular Modelling

There are various uses of DNA molecular modelling in Genomics and Biotechnology research applications, from DNA repair to PCR and DNA nanostructures. Two-dimensional DNA junction arrays have been visualized by Atomic force microscopy. DNA molecular modelling has various uses in genomics and biotechnology, with research applications ranging from DNA repair to PCR and DNA nanostructures. These include computer molecular models of molecules as varied as RNA polymerase, an E. coli, bacterial DNA primase template suggesting very complex dynamics at the interfaces between the enzymes and the DNA template, and molecular models of the mutagenic, chemical interaction of potent carcinogen molecules with DNA. These are all represented in the gallery below.

Technological application include a DNA biochip and DNA nanostructures designed for DNA computing and other dynamic applications of DNA nanotechnology. The image at right is of self-assembled DNA nanostructures. The DNA "tile" structure in this image consists of four branched junctions oriented at 90° angles. Each tile consists of nine DNA oligonucleotides as shown; such tiles serve as the primary "building block" for the assembly of the DNA nanogrids shown in the AFM micrograph. Quadruplex DNA may be involved in certain cancers. Images of quadruplex DNA are in the gallery below.

Nucleic Acid Structure

The sequence or primary structure of a nucleic acid is the exact specification of its atomic composition and the chemical bonds connecting those atoms. As nucleic acids, e.g. DNA and RNA, are unbranched polymers, this is equivalent to specifying exact sequence of nucleotides that comprise the whole molecule. This sequence is written as a succession of letters representing a real or hypothetical DNA molecule or strand. By convention, the primary structure of a DNA or RNA molecule is reported from the 5' end to the 3' end.

The sequence has capacity to carry information. When used in reference to biological DNA, which carries the information which directs the functions of living beings, the term genetic sequence is often used. Sequences can be read from the biological raw material through DNA sequencing methods. Primary structure is sometimes mistakenly termed *primary sequence*, but there is no such term, as well as no parallel concept of secondary or tertiary sequence.

Nucleotide

Nucleotides are molecules that, when joined together, make up the structural units of RNA and DNA. In addition, nucleotides play central roles in metabolism. In that capacity, they serve as sources of chemical energy (adenosine triphosphate and guanosine triphosphate), participate in cellular signalling (cyclic guanosine monophosphate and cyclic adenosine monophosphate), and are incorporated into important cofactors of enzymatic reactions (coenzyme A, flavin adenine dinucleotide, flavin mononucleotide, and nicotinamide adenine dinucleotide phosphate).

Nucleotide Structure

A nucleotide is composed of a nucleobase (nitrogenous base), a five-carbon sugar (either ribose or 2'-deoxyribose), and one to three phosphate groups. Together, the nucleobase and sugar comprise a nucleoside.

The phosphate groups form bonds with either the 2, 3, or 5-carbon of the sugar, with the 5-carbon site most common. Cyclic nucleotides form when the phosphate group is bound to two of the sugar's hydroxyl groups. Ribonucleotides are nucleotides where the sugar is ribose, and deoxyribonucleotides contain the sugar deoxyribose. Nucleotides can contain either a purine or a pyrimidine base.

Nucleic acids are polymeric macromolecules made from nucleotide monomers. In DNA, the purine bases are adenine and guanine, while the pyrimidines are thymine and cytosine. RNA uses uracil in place of thymine. also nucleotide plays an important role in thinking.

Synthesis6

Nucleotides can be synthesized by a variety of means both in vitro and in vivo. In vivo, nucleotides can be synthesised de novo or recycled through salvage pathways. Nucleotides undergo breakdown such that useful parts can be reused in synthesis reactions to create new nucleotides. In vitro, protecting groups may be used during laboratory production of nucleotides. A purified nucleoside is protected to create a phosphoramidite, which can then be used to obtain analogues not found in nature and/or to synthesize an oligonucleotide.

Bibliography

Ayala, F. J.: *The Genetic Structure of Populations*, W.H. Freeman & Co., San Francisco, California, 1977.

Barnaby, W.: *Plague Makers: The Secret World of Biological Warfare*, Vision Paperbacks, London, 1999.

Berg, Paul, and Singer, Maxine: *Dealing with Genes: The Language of Heredity. Mill Valley*, University Science Books, CA, 1992.

Berry, R. J.: *Inheritance and Natural History*: Collins, London, 1977.

Bradley, W.G.: *Neurology in Clinical Practice*, Butterworth Heinemann, Boston, 2000.

Crick, F.: *Life Itself: Its Origin and Nature*, W.W. Norton, New York, 1982.

Curtis, M.: *The Geometry of DNA: A Structural Revision*, Blue Gallery, London, 1996.

Dawkins, R.: *The Selfish Gene*, Oxford University Press, New York, 1976.

Dunbar, Robert E.: *Heredity*, Franklin Watts Publisher, New York, 1978.

Fisher, R. A.: *The Genetical Theory of Natural Selection*, Clarendon Press, Oxford, 1930.

Glover, D. M., and Hames, B. D.: *Genes and Embryos*, Oxford University Press, New York, 1989.

Goldman, L.; Ausiello, D. A.: *Cecil Textbook of Medicine*, Saunders, Philadelphia, 2004.

Hamerton, J. L.: *Human Cytogenetics*, Academic Press, New York, 1971.

Hanley, R.: *Is Data Human? The Metaphysics of Star Trek*, Boxtree, London, 1998.

Jonoska, N.: *Self-Assembling DNA Graphs, DNA-Based Computers VIII*, Springer-Verlag, Berlin, 2003.

Kelves, D.: *In the Name of Eugenics: Genetics and the Uses of Human Heredity*. Harmondsworth: Penguin.

Klein, Aaron E.: *Threads of Life: Genetics from Aristotle to DNA*, The Natural History Press, Garden City, New York, 1955.

Lewin B.: *Genes VII*, Oxford University Press Inc., New York, USA, 2000.

Lewontin, R.C.: *The Doctrine of DNA: Biology as Ideology*, Penguin, Harmondsworth, 1993.

Lyon, Jeff and Gorner, Peter. Altered Fates: *Gene Therapy and the Retooling of Human Life*, W. W. Norton and Company, New York, 1995.

Margulis, L.: *Symbiosis in Cell Evolution,* W.H. Freeman, San Francisco, 1981.

Nelkin, D. and M.S. Lindee: *The DNA Mystique: The Gene as a Cultural Icon*, W.H. Freeman, New York, 1995.

Rietman, Ed.: *Molecular Engineering of Nanosystems,* Springer, New York, 2001.

Stebbins, G. L.: *Darwin to DNA, Molecules to Humanity*, W. H. Freeman, San Francisco, 1982.

Sturtevant, Alfred: *History of Genetics*, Harper and Row, New York, 1965.

Suzuki, D., and Knudtson, P.: *Genethics: The Clash Between the New Genetics and Human Values*, Harvard University Press, Cambridge, Mass, 1989.

Watson, J. D.: *The Double Helix*, Antheneum, New York, 1968.

Williams, J.G. and R.K. Patient: *Genetic Engineering*, IRL Press, Oxford, England, 1988.

Winchester, A. M.: *Heredity, Evolution and Humankind*, West Publishing Co., St. Paul, Minn., 1976.

Winston, R.: *The Future of Genetic Manipulation*, Phoenix, London, 1997.

Wright, L.: *Twins: Genes, Environment and the Mystery of Human Identity*, Weidenfeld & Nicolson, London, 1997.

Index

❑❑❑